Embryology and Genetics

THOMAS HUNT MORGAN
Professor of Biology
California Institute of Technology
California

AGROBIOS (INDIA)

Published by:
AGROBIOS (INDIA)
Agro House, Behind Nasrani Cinema
Chopasani Road, Jodhpur 342 002
Phone: 91-0291-2643993, 2643994, Fax: 2642319
E. mail: agrobiosindia@gmail.com, info@agrobiosindia.com
Website: www.agrobiosindia.com

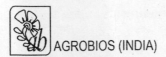 AGROBIOS (INDIA)

Price I $12.94 $12.94
ISBN: 978-81-7754-076-5

Published by: Dr. Updesh Purohit for Agrobios (India), Jodhpur
Lasertypeset at: Yashee Computers and Printers, Jodhpur
Cover Design by: Reena
Printed at: Bharat Printing Press, Jodhpur

PREFACE

It has always seemed to me strange that, while the story of evolution arouses widespread popular interest, the evolution of an individual from an egg excites so little curiosity. Perhaps the miraculous origin of man and the mysteries surrounding his appearance on the earth, made familiar in the folklore of many peoples, sufficiently account for the interest of those in whose childhood these stories were imbibed, while the prosaic fact that a chick develops from the egg of a hen is accepted as a commonplace rather than as a fascinating problem.

If the origin of life on the earth, its evolution culminating in the races of mankind, seems to call for a miracle, then the same or at least another miracle seems called for each time an egg develops into an adult organism. If we could imagine that an observer with the limited capacities of a man had been present throughout the millions of years during which the evolution of plants and animals was taking place, it is doubtful whether he would have been any wiser as to what was going on around him than is a palaeontologist of today. A fragment of the evolution of the higher groups is, in fact, preserved in the rocks of the earth's surface. Their bones and shells, arranged in their probable sequence, furnish a picture of what took place. Only a small fraction of what went on is preserved, it is true, yet enough to furnish clues as to the sequence of events. But the attempts of palaeontologists to deduce the causal factors at work have led at most to inconsequential conclusions, not much better than the attempts of primitive races to explain the physics of an eclipse of the sun or the physiological action of herbs.

The student of embryology who observes the changes of form that take place as the egg passes through its embryonic stages molding itself into the likeness of the parents from which it came, is, in reality, in little better position than the fictitious individual present during the evolution of the organisms living today; for, like the latter, he sees next to nothing of what is behind the changes he observes. His speculative adventures have not succeeded in bringing the phenomena of development within the framework of exact science. Only when he comes to apply the method by which science separates the wheat from the chaff, namely, the use of working hypotheses controlled by quantitative measurements, or in a word experimental methods, does he begin to make embryology an exact science.

The appeal of modern embryologists to the experimental method is not so much with the intention of discovering how far chemistry and physics enable him to explain what is going on behind the scenes as it is to utilize the experimental method as a tool to discover the nature of the visible phenomena of development. If in the following pages a good deal of attention is given to the purely visible aspects of development, it is done only in so far as is necessary to make clear the nature of the experiments designed to carry the problems a step further. The experimental study of development is still in its infancy, but enough has been found to give encouragement for further work.

Since 1900, when the discovery of Mendel s work became known, one of the most amazing developments in the whole history of biology has taken place. The fundamental laws of heredity are now known, and since it is through the egg that the hereditary properties of the individual are carried on from generation to generation, the importance of an understanding of development to supplement the knowledge of the laws of heredity is apparent, and the interlocking of these two experimental branches of biology has become a subject of absorbing interest. In the follow_ _ _g ᵀ have attempted to point out in a simple way their interrelations. That much remains to be done will be only too obvious, but with the openings furnished by

PREFACE

the experimental investigation of heredity and embryology there is promise that a great deal more is within our reach.

By introducing into the text many figures, I have let them take the place, as far as possible, of verbal description. No pains have been spared to make the diagrams simple and clear. Descriptive embryology is, in fact, a series of pictures of the stages through which the embryo passes from the egg to the form of the adult. It is with some hesitation and even regret that I finally decided to omit from the text the names of living authors from whose works most of the statements there made have been drawn. In so brief an account of the present interrelations of embryology and genetics, the text would be dotted everywhere with references which do not seem to be essential in a book of this kind. The omissions are rectified, I hope, to some extent by giving credit in the legends of the figures to the authors from whose publications the illustrations have been taken. For students of embryology there has been added at the end a bibliography under chapter headings which, while it does not attempt to give the whole literature of the subject, does give the titles of the papers from which many of the statements in the text are taken. Those who wish to go further into the fields of experimental embryology and of genetics will find the references to these two subjects in the "General References" preceding the literature of the separate chapters. Several friends and colleagues have kindly read the chapters in their immediate fields of work, and I am much indebted to them for suggestions and corrections. To Dr. Albert Tyler and to my wife, L. V. Morgan, who have read the entire text, I am especially indebted.

Pasadena, California   THOMAS HUNT MORGAN

CONTENTS

EMBRYOLOGY AND GENETICS

CHAPTER I
INTRODUCTION

The stratified rocks of the earth's surface reveal the most recent part of the long history of the evolution of the animals and plants living at the present time. While it took millions of years to bring about these changes, the development of each individual from an apparently simple egg to the visibly complex form of the adult is now only a matter of days, or even hours. The comparison may be misleading, however, since there have probably been long periods when little or no change took place in the species, and the next advance, appearing in a single individual, may actually have occurred in infinitesimal time, from gene to gene, involving only a sudden alteration in one of the units of heredity.

The identification of the egg cells with the single-celled ancestors from which the higher forms have evolved calls for qualification. The converse statement may be nearer the truth, namely, that the egg of today is as different from the original unicellular ancestor as the adult today is different from that ancestral adult. Both statements call for reservations, for everything turns on what is meant by likeness and difference. In the egg there are all the potentialities for quickly developing the characteristics of the adult form, and in this sense the egg differs immensely from the original one-celled ancestor. The difference lies in the units of heredity in the two cases: only in their visible form are the protozoön and the egg somewhat alike Since we know nothing about the constitutional differences between the hereditary elements in the original protozoön and those of the egg of today, it is futile to attempt to make any serious comparisons between the relative complexity of the two. Only superficially are they alike in their visible structures

In another respect, however, we may make comparisons. The ancestral type needed to pass through fewer visible changes from egg to adult. In the unicellular forms, the protozoa, that multiply by self-division each daughter cell has little more to do than to enlarge to the original size, and in the lower metazoa the stages, after division of the egg, are very few compared with those of higher forms. But even then the comparison may be misleading, for in the higher forms it is the visible changes that are considered, and we think of them most often as changes in form or structure, while the physiological processes in the unicellular and multicellular types are probably much more alike. In the higher forms these processes are separated into organ systems, but they may be much the same as in the protozoa. Descriptive embryology concerned itself entirely with changes in form, and very little with the physiology of development. Only recently has the latter received serious attention, although there have always been a few students interested in the physiology, especially in the later stages, of the vertebrate embryo.

For many years—let us say between 1850 and 1900—embryologists were engrossed with the idea that development of higher forms recapitulated the entire historical path over which their evolution had passed. This became known as the recapitulation theory. An immense amount of purely descriptive embryological work was carried out under the influence of this theory, and today the embryology of all types of animals is known, often in the most minute detail. In small transparent eggs the developmental stages may be followed under the microscope; and, even in eggs that are more opaque, technical methods have been devised that reveal the changes taking place beneath the surface. The perfection of these methods—staining, imbedding in paraffin, cutting into thin slices, mounting these in balsam on glass slides, and reconstructing the whole in wax—occupied for a long time the attention of a great number of professional embryologists, to the exclusion of considerations dealing with the physical and chemical events that lie behind these visible stages of development. The historical appeal was irresistible, especially if one believed that what he was seeing and describing was the history of "creation"—or, as it was called, evolution. There was soon established an immense body of information concerning the development of all the main animal forms. Accurate observation was called for, of the same order as that of all pictorial art. Beautiful illustrations of the development from egg to embryo appeared in a host of monographs. The better the artist, the more brilliant his performance. The anatomy of development became as well known as the older anatomy of adult structures that had likewise called for close observation and an artistic sense of representation in color and perspective.

During the final years of the last century and down to the present time a new interest appeared, called experimental embryology, and sometimes developmental mechanics. The reaction that had set in against the old interpretation of the developmental stages as a recapitulation of the ancestry was in part responsible for this change of interest. New ways of finding out something of what is going on behind the scene, the discovery of potentialities in the egg never before suspected, the application of methods to bring about unnatural changes in the development the emphasis on the rôle of the environment in normal development, all conspired to awaken new interest.

Into the new fields of exploration many of the young embryologists entered with renewed enthusiasm. A great deal was revealed and many more problems, very different in kind from those that had fascinated the preceding generation, appeared. Here it seemed was the possibility of further advance in an understanding of the developmental processes; and the idea that embryology could be placed on an experimental basis was especially attractive to those who were familiar with the great advances that the experimental method in chemistry and physics had brought about. The embryologist found himself dealing with problems so different that it did not seem possible to apply at once the laws of chemistry and physics. He dealt with such complex materials as proteins, colloids, and with such complex problems as surface forces, permeability, etc., that the physical scientists themselves had not yet brought into line with the rest of their work. In fact, nearly all of the experimental work, so called, in embryology remained still on the biological level. It made known many conditions in the development of the egg that had never before been suspected, but the appeal to physics and chemistry of the so-called developmental mechanics was more often by analogy than by demonstration, and even "chemical embryology" has been largely a description of the kind of chemical compounds found in the egg and embryo. It is true that the transformation of some of these compounds into the other substances or into the finished product is an essential part of the embryological problem, but the embryologist is very largely concerned with the kinds of reactions that lead to the particular changes in form of the embryo, as well as with the origin of substances from other materials.

The extraordinary fact that an egg with little *visible* organization develops into a complicated adult, with a vast amount of organization, had aroused the interest of the philosophers from Aristotle to Whitehead, and in a broad way they realized the mystery of something happening that had no parallel in other fields of scientific interest. These thinkers were mainly impressed with the kind of organization expressed in form as the most important feature of development, and today this still remains as the most outstanding feature of development. That these changes in form might depend on chemical changes in the embryo was either taken for granted or ignored.

The most discussed "principle" of philosophy goes under the name of entele hy. The entelechy, supposedly the same idea under that name in Aristotle's teachings, was postulated as a principle, guiding the development toward a directed end—something beyond and independent of the chemical and physical properties of the materials of the egg; something that without affecting the energy changes directed or regulated such changes, much as human intelligence might control the running or construction of a machine. The acceptance of such a principle would seem to make it hardly worth while to use the experimental method to study development, since it would be directed and regulated by the entelechy. In fact, the more recent doctrine of "the organism as a whole" is not very different from the doctrine of entelechy, except in so far as other ways, by which the whole might be coördinated in an ultra or supermaterialistic way, might be imagined.

Therefore, unless it be granted that the principles involved in development are of a different order from physical principles in the broadest and most recent usage of this term, it would seem better to table these metaphysical questions, and to try to discover, despite the amount of time and labor involved, how far a knowledge of the chemical and physical changes taking place in the egg will carry us toward an understanding of the developmental processes. It may, of course, be found that an understanding of the kind of system present in the egg, sometimes still called the organization of the egg, will require relatively new principles peculiar to colloid systems, balanced salt solutions, semipermeable membranes, phase boundaries, etc.; but if these "principles" are still found to follow physical and chemical laws, whether the old ones or new ones, for large-scale phenomena with which embryology appears to be concerned, the study of embryology would still come to range itself under a broader conception of natural processes, including in its scope both living and dead material. If, on the other hand, it should turn out that an understanding of living materials calls for something quite new to the physical sciences, it will then be time to examine the nature or un-nature of this something. Meanwhile it seems clear that the next step should be a determined effort to learn all that we can about the kind of system or configuration that constitutes the egg. This statement does not mean that we should resort entirely to the kind of analyses which chemists and physicists have invented for the study of their kinds of materials, but that we should not neglect any possible means of penetrating further by experimental methods, on the biological level, into the behavior of such systems.

It is unsafe to say that the physico-chemical problems are different from the biological problem until we know more about the latter. For it must be obvious to every student of embryology that we have only begun to get information as to the "organization" of the egg on the biological level, and know as yet very little about the chemistry and physics of development. Should it turn out that neither the classical mechanics, nor the new physics suffice, the ground will at least be prepared for the discovery of some new kinds of principles that apply to living things. But until it has been shown that what we call the property or properties of living things are entirely out of line with what is known as non-living systems, it may be shortsighted to resort to obviously metaphysical principles, or even to temporize with them. It is this alternative that separates those whom the philosophers insist on calling mechanists, and those whom the biologists call metaphysicians. There is no need to attempt a compromise by saying that each has his own realm, because the scientist regards mysticism as an outmoded way of attempting to offer a finalistic solution of the problems he studies.

Most modern biologists are not, however, so much impressed by the idea that there is a principle of life as they are by the great variety of phenomena shown by living things. It seems to them premature as well as pretentious to discuss some imaginary ideal property of life when there is abundant evidence pointing to the conclusion that there are many properties of living things of many different kinds, each crying out for solution before attempting to synthesize them into life. Of course one may pick out one or more of these, such as consciousness, or purpose, or free will, and make it the *sine qua non* of living things, but it should not pass unnoticed that the selection is usually one of the most obscure phenomena of living things.

Formative forces, polarity, symmetry, and purposeful regulations are examples of this in the embryological realm.

The story of genetics has become so interwoven with that of experimental embryology that the two can now to some extent be told as a single story. It is true there are still wanting many important points of contact, but enough is known co make it possible to attempt to weave them together into a single narrative. Although each has developed in large part independently of the other, nevertheless today their interdependence is so obvious that the geneticist takes for granted the main outlines of the facts of embryology, and the embryologist is coming to realize his dependence on the evidence from genetics. For example, cell division and the behavior of the chromosomes at maturation of the eggs and sperm have supplied the working scheme for the theory of heredity. The changes that take place during the maturation of eggs and sperm are contributions from embryology. Conversely, genetic analysis has made it possible to go behind these visible changes into the very constitution of the chromosomes themselves. The common meeting point of embryology and genetics is found in the relation between the hereditary units in the chromosomes, the genes, and the protoplasm of the cell where the influence of the genes comes to visible expression. Concerning the manner of functioning of the genes during development, I have contrasted, in the following pages whenever an opportunity arises, two possible views, and suggested a third. The implication in most genetic interpretation is that all the genes are acting all the time in the same way. This would leave unexplained why some cells of the embryo develop in one way some in another, if the genes are the only agents in the results. An alternative view would be to assume that different batteries of genes come into action as development proceeds. The former view, namely, that all the genes are acting all the time in the same way, leaves the embryological problem where it has always been supposed to be, viz., in the protoplasm. The alternative view might appear to give a formal explanation of development, but is inconsistent with results obtained by changing the sequence of the cleavage planes by compression. Roux and Weismann attempted to explain development in somewhat this way, by assuming that the determinants in the chromosomes are qualitatively sorted out during development. There was at the time no evidence in favor of this view, and there is now much that is opposed to it. The idea that different sets of genes come into action at different times is exposed to serious criticism, unless some reason can be given for the time relation of their unfolding.

The following suggestion may meet these objections. It is known that the protoplasm of different parts of the egg is somewhat different, and that the differences become more conspicuous as the cleavage proceeds, owing to the movements of materials that then take place. From the protoplasm are derived the materials for the growth of the chromatin and for the substances manufactured by the genes. The initial differences in the protoplasmic regions may be supposed to affect the activity of the genes. The genes will then in turn affect the protoplasm, which will start a new series of reciprocal reactions. In this way we can picture to ourselves the gradual elaboration and differentiation of the various regions of the embryo.

CHAPTER II

DEVELOPMENT AND GENETICS

It must have been known, long before there is any recorded human history, that some animals develop from eggs, at least those animals whose eggs are visible to the unaided eye. Birds, lizards, frogs and fish have large eggs; most other animals have smaller eggs, many of them just visible to the naked eye, Fig. 1. The human egg measures only one-fifth of a millimeter, and since it develops inside the mother and, therefore, out of sight, the origin of the human embryo remained a mystery until von Baer identified the mammalian egg in 1820.

The rôle of the male in procreation was much more difficult to discover. The need of the semen to start the development of the egg may have been suspected in very early times, but the essential elements in the semen could not have been imagined until the presence of animalculae or spermatozoa in it had been discovered by Hamm and Loewenhoek in 1677. That these animalculae were the real agents in development was not generally accepted for a long time. Only after 177 years was the actual entrance of the sperm into the egg observed by Newport (1854). By this time it was recognized that the egg is a cell, and that the spermatozoön is also a cell.

The next advance was made by Hertwig in 1875, who observed that the sperm head, after entering the egg, enlarges to become a recognizable nucleus which fuses with the nucleus of the egg. The two combined nuclei then give rise to the nucleus of the dividing egg, which is the forerunner of all the nuclei of the cells of the resulting embryo. Since it was believed that the child inherits equally from the two parents, and since the spermatozoön contributes only its nucleus to the combination—its tail not entering the egg as a rule—

it was concluded that the hereditary materials are contained in the nuclei.

At the time when this inference was drawn, it was known that each time the nucleus prepares for division of the cell, rod-shaped or v-shaped bodies appear, having a strong affinity for certain dyes; hence called chromosomes. Their number, size and shape are characteristic of each species. The way in which this constancy is maintained was explained when Van Beneden (1883) made the discovery that the sperm nucleus brings into the egg half the typical number of chromosomes, the paternal chromosomes; and that the egg nucleus also contains, before fertilization, the half number, the maternal chromosomes. In other words, the mature egg and spermatozoön

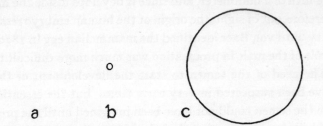

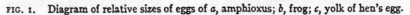

FIG. 1.   Diagram of relative sizes of eggs of *a*, amphioxus; *b*, frog; *c*, yolk of hen's egg.

each contains exactly half of the original number of chromosomes characteristic of the species. The way in which the reduction in number of chromosomes of the mature germ cell is brought about was discovered later by a study of the changes that take place in the eggs and sperm cells in their final stages of development. These changes will be described in the next chapter.

It has been shown by an analysis of the transmission of the characters of the parents that whole chromosomes are not the ultimate units of inheritance, but carry within themselves smaller units called genes. These are arranged in each chromosome in linear order —like beads on a string. The evidence that the genes are the ultimate units of heredity does not rest on direct observation, since the genes are beyond the limits of vision of our microscopes, but by inference from the facts of inheritance.

In this brief summary of the steps that have led to the modern conception of the mechanism of heredity, there should be included another contributory line of investigation that helped to clear up the rôle of chromosomes in heredity. This evidence came from experimental embryology. If an egg is cut in two before fertilization, each piece may be fertilized by a single sperm, and each may develop into an embryo. One of the pieces contains the original egg nucleus. Its development follows that of the whole egg; it contains, after fertilization, the full number of chromosomes, the diploid number. The other piece is without a nucleus at first, but, after a sperm enters, it comes to contain a sperm nucleus with the half number of chromosomes, one of each kind. This number is the haploid number. The piece develops, and its development establishes the fact that a single set of chromosomes suffices to give normal development. It is true that these "haploid" embryos are sometimes weak. In some cases at least, it has been shown that some of the embryos, that start with the half number of chromosomes and develop as far as the adult stages, have doubled the number of chromosomes by suppressing one of the first divisions of the protoplasm of the egg. Nevertheless, the fact that a typical embryo may also develop from the haploid piece demonstrated that one set of chromosomes suffices to produce the characteristics of the individual. There is further evidence from parthenogenetic development in support of this conclusion.

Other evidence, as to the significance of the chromosomes, was furnished by an ingenious experiment of Boveri. It was known that when two spermatozoa enter the same egg of the sea urchin, the first division is into three or four cells instead of into two as in the normal development. Such eggs do not develop normally. As had been earlier shown by Hertwig the first division of the chromosomes in these dispermic eggs is irregular, and the three, or four resulting cells get different numbers of chromosomes. It was also known at this time from the result of some experiments by Driesch that if the first two cells of the sea urchin's egg are separated, each will form a normal embryo and similarly for the four cells. Each contains, of course, the usual double set of chromosomes. Boveri found, on the contrary, that if the three cells of dispermic eggs are separated, only rarely

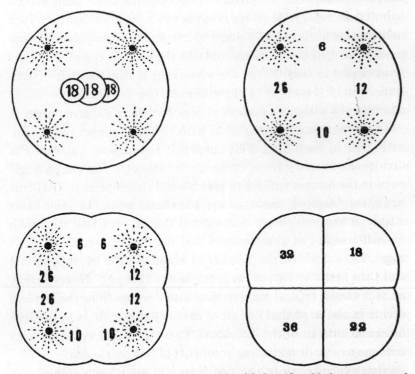

FIG. 2.   Diagram of dispermic egg of sea urchin with 3 × 18 — 54 chromosomes and four asters (above to left). The other figures show one of the many possible distrib tions of the chromosomes to the four resulting cells. (After Boveri.)

does one of them develop into a normal embryo, Fig. 2. He concluded that the failure of most of these isolated cells to give normal development is due to the irregular distribution of the chromosomes, and that the only cells that develop into normal embryos are those containing, by chance, at least one full set of chromosomes. This inference is partly true, but not the whole truth. It is true that one set alone suffices, but if there are other chromosomes also present, the course of development is upset. Under the conditions of the experiment it could only rarely happen that just one set of chromosomes and no more would pass into one of the three cells. A more complete analysis of the way in which the chromosomes are distributed in such eggs would be necessary to make plausible the numerical relations on which Boveri based his conclusions, since any number over one set is a disturbing factor. Later, from genetic sources, it has been shown that the basal set may be multiplied three, four or more times and still give normal development; but the addition of a single chromosome or two to a set, may be injurious.

The most general idea concerning the gene is that it is an entity with two fundamental properties. First, its power to grow and divide; second, its power to bring about changes in the protoplasm outside the nucleus—changes that affect the chemical and physical activities of the protoplasm. The first of these attributes rests on the visible division of the chromosomes, which split lengthwise at each division—each half containing all the properties of the original chromosomes. In other words, the gene always divides quantitatively, and then grows to the size of the original gene. The same property must reside in each new gene that perpetuates itself. The power to divide is also a property of the whole cell, but with the important difference that at some of the divisions of the egg one of the daughter cells may contain materials to some extent different from those of its sister cell. This inequality of the early divisions of the egg cell may furnish a clue, as pointed out in the last chapter, as to the way in which different regions of the segmenting egg come to develop in different ways.

The second attribute of the gene has no direct observational basis, but rests on a supposedly logical deduction from the results of genetic analysis. This genetic evidence shows that when a gene mutates, without losing its property of self-perpetuation, it brings about changes in the character of the resulting individual. The argument, in fact, is based on the reversed relation of gene to character, for what we see is the appearance of a new character, or set of characters, and then by analysis we refer this change of character to a change in a gene. What is significant here is that the change can be analytically traced to a particular spot or locus in one of the chromosomes—hence to a single gene. Granting, then, that we are safe in inferring that the change in character depends on some property of the new gene, the question arises as to how the gene produces its effect on the protoplasm of the cells, for it is in the protoplasm that the character is manifest.

It is conceivable that the effect might be due to some dynamic action of the gene on the surrounding protoplasm. This possibility cannot at present be proved or disproved, but since many or most changes in cells are chemical in nature, it seems more plausible to assume that the gene sets free some chemical material—perhaps in the nature of a catalyst—that induces certain chemical changes in the protoplasm.

There are other questions to be considered before going further into the relation of genes to characters. It is known that when a gene mutates it produces changes throughout the organism. Some of the changes are great enough to be plainly visible; others less so, or even so small as to escape visible detection, being seen only in their physiological aspects, such as the death or the length of life of the mutant. In the early days of genetics, i.e., at the beginning of the century, "unit characters" were supposed to furnish the basis for genetic work, and by inference each gene was supposed to produce a specific effect in only one character at a time. This premature inference was very soon found to be erroneous when the manifold effects of each genic change came to be known. It is true that in most genetic work one particular character is selected as the symbol of the gene concerned with its appearance, but this selection is only because the character selected is the one most easily identified, or one that is less variable—i.e., less affected by the environment.

The next point that calls for consideration is that each character of the adult is the product of many genes, or it may even be said of all the genes if the whole history of the affected organ is retraced to the egg.

The meaning of genic balance is intimately related to these questions. It has already been mentioned that the embryological evidence has shown that one set of genes at least is necessary for normal development; two sets is the most usual condition. Many cases are known where four, six, eight or more sets also give normal results. This is part of the evidence on which the idea of genic balance rests —an idea that was originally, if rather vaguely, implied in the embryological evidence that has been given. The idea has come, however, to have a much more definite meaning when the individual genes are taken into account, and its earlier speculative character has given way, in the light of more recent experimental genetic data, to a more positive formulation.

The central idea of genic balance is that all the genes are acting, and what is produced is the sum total of their influence. If only one gene is changed, i.e., mutates, the product is changed in some degree, certain organs being more affected than others; but still all the genes are concerned. In other words, the new gene acts only as a differential. This formulation gives a consistent picture of the end-products of the genes, but it is quite inadequate to explain the sequence of changes through which the embryo passes.

CHAPTER III

THE EGG AND THE SPERMATOZOÖN

Neither the large eggs of some animals nor the small eggs of others furnish any visible evidence of the kind of individual to which they will give rise. While it is true that each kind of egg has characteristics by which the species to which it belongs can be identified, these visible differences are secondary, such as the kind of membrane surrounding the egg, the presence of pigment, or the amount of yolk, or the size of the egg. Many animals are bilaterally symmetrical, but only with rare exceptions, as in the case of many insects, do the unfertilized eggs show evidence of bilaterality.

THE MATURE EGG

Most eggs are spherical; many of them have a polar field that marks a definite region to which all of the early changes are related. The materials of the egg are stratified with respect to the pole. There is more protoplasm in the polar region, while in the opposite hemisphere there is more yolk, Fig. 3. Looked at from the pole, the materials are radially symmetrical with respect to the pole. There are no indications as to where the planes of symmetry of the embryo will come to lie.

The nucleus of the ripe ovarian egg is exceptionally large, Fig. 3, in comparison with the nuclei of most other cells. It lies usually in an excentric position with respect to the center of the egg, somewhat nearer the pole. It is filled with a semifluid sap, through which there is a fine network of chromatin material. The fluid in the nucleus is separated from the surrounding protoplasm by a denser wall of protoplasm.

When the egg is ready to be set free from the ovary, conjugation of the chromosomes in pairs has taken place. The wall of the nucleus is dissolved, and its fluid sap diffuses through the protoplasm of the rest of the egg. The paired chromosomes now appear, Fig. 4, and become arranged on the first polar spindle, Fig. 4a. There appear to be only half as many chromosomes as there were original chromosomes in the earlier divisions of the egg cells, since each chromosome consists of two united chromosomes.

The chromosomes attach themselves to the spindle fibers at the equator of the spindle at a definite point on each chromosome, known as the attachment point, which may be in the middle, or at or

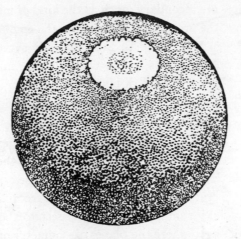

FIG. 3.   Ripe ovarian egg of frog.

near the end of the chromosome. This attachment point is an important organ of the chromosome. Should the attachment point be lost, as sometimes happens when a chromosome becomes broken, the fragment without it fails to become attached to the spindle. In consequence it will sooner or later become lost.

The spindle with the chromosomes now moves toward the pole of the egg, and assumes there a radial position. The protoplasm at the pole begins to protrude, and one pole of the spindle moves into the protrusion, Fig. 4b. A chromosome from each pair passes to the outer pole, and its mate to the inner pole. The protruding protoplasm pinches off, Fig. 4c. It contains the outer group of chromosomes. The inner group remains in the egg The first polar body has been given off.

Around the chromosomes that are left in the egg a new spindle developes, which soon takes a radial position beneath the pole, Fig. 4d. The chromosomes that had remained in the egg after the first polar body was given off, are seen to be split at this time; half of each goes out into the second polar body, half stays in the egg, Fig.

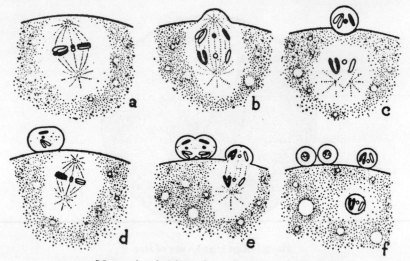

FIG. 4.   Maturation divisions of egg; formation of polar bodies.

4e. While this is taking place the first polar body has also divided into equal parts. The outcome is four cells, Fig. 4f, namely, the three polar bodies and the egg, which have resulted from the two cell divisions, one of which differs from all other divisions in that the chromosomes of each pair have been separated.

Many lower marine animals, jellyfish, sea urchins, worms, oysters and fish, set free their sperm in the sea water, generally at the same time that the eggs are set free. There is an immense loss of spermatozoa, but they are so numerous that all of the eggs become fertilized. In a few cases, as in many of the salamanders, the male deposits packets of semen on the bottom of the pool to which the two sexes resort at breeding time. The female crawls over the packets and takes them into the cloaca. In the oviduct the packets are dissolved as the sperms are set free to fertilize there the eggs.

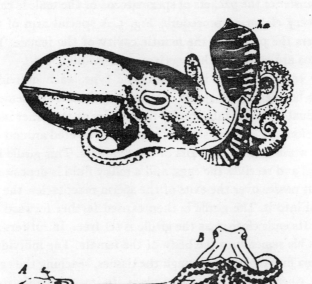

FIG. 5. Octopus, male, showing hectocotyl arm (*ha*). Copulation (below), small male, *A*; large female, *B*.

The male frog clasps the female, riding on her back, and sets free his semen over the eggs as they emerge from the oviduct. In birds and lizards the male and female copulate by bringing the cloacal openings into contact; the semen is shot into the oviduct of the female, where fertilization takes place. In the mammals a copulative organ, the penis, is present in the male. The sperm duct traverses its length. This organ is introduced into the vagina of the female where the semen is deposited. Through the swimming activity of the spermatozoa themselves, or by peristaltic contractions of the oviducts, they pass or are carried through the length of the oviduct and fertilize the eggs in the upper reaches of the tubes.

Snails also possess copulatory organs through which the semen is conveyed to the oviduct of the female. In the squid and cuttlefish the transfer of the packets of spermatozoa of the male is carried out by a very elaborate procedure, Fig. 5. A special arm of the male transfers the packet to the mantle cavity of the female. The spermatozoa are set free as the eggs leave the oviduct.

The earthworm is a hermaphroditic type, each individual producing both eggs and sperm. During copulation the semen of one individual is passed into special receptacles of the other individual. Just before the eggs are set free, a girdle is secreted around the body of the worm behind the exits of the oviducts. This girdle is moved forward and receives the eggs, and a milky fluid is deposited within it. As it passes over the exits of the sperm receptacles, the semen is ejected into it. The girdle is then carried farther forward over the head. Its ends close in as the girdle is set free. In rotifers the male injects his semen into the body of the female. The individual spermatozoa find their way through the tissues, reaching the eggs in the ovary, where fertilization takes place. In insects highly specialized copulatory organs are present in both sexes. During copulation the semen passes through the penis of the male into the oviduct of the female, and finds its way into her seminal receptacle where it is stored. As each egg passes by the opening of the receptacle, a few sperm are passed out, one or more entering each egg. The spermatozoa may remain alive in the receptacle of the female for days, months, or even years. The queen bee, for example, which is fertilized only once in her life, retains enough living sperm to fertilize her eggs for the rest of her life, which may last several years.

Hundreds, and sometimes many thousands, of eggs are produced in the ovaries of the female. They ripen singly or all at one time. The young ovarian eggs increase in number by the ordinary processes of cell division, and only after passing through many divisions do they undergo the final stage or maturation stages in which the number of chromosomes is halved.

The eggs in most animals mature at certain seasons of the year, when all the eggs that are ripe are set free. Animals with large eggs usually lay fewer than those with small eggs. Some birds lay only two eggs, others a dozen, or more, each year. Frogs lay several hundreds of eggs. Fish, and many lower animals, set free larger numbers of eggs, especially those that set free their eggs in the sea. It has been estimated that the cod produces 6,652,000; the oyster 60,000,000; a snail (aplysia) 2,000,000. The queen bee is said to produce several million eggs; they are laid one at a time in the cells of the comb during the five to eight years of her life.

In mammals the number of eggs set free from the ovary is relatively small. The young born per year are usually one in cattle and horses, and up to a dozen in rodents and swine. Some of the smaller mammals, such as mice and rats, may produce several litters a year, especially if well fed and kept warm. In the higher apes and man, where the single embryo is retained for a long time in the uterus of the mother, the total output is relatively very small, even although each egg is microscopic in size. In the course of the life of a woman about 400 eggs may be set free from the ovaries and pass into the oviducts, but few of them ever become fertilized.

The abundance of a species does not depend so much on the number of eggs produced as on the chance that the embryo reaches the adult stage. Hence animals that protect their young, by retaining them in the mother's body, or by care after birth, may, in the long run, leave as many descendents as do those producing many more eggs.

THE RIPE SPERMATOZOA

The typical spermatozoön is a minute, threadlike "animalcule" with a rounded head at one end, and a long tail, Figs. 6a–f. Only with very high powers of the microscope is it possible to make out its structure. When set free in water the "sperm" swims about in a

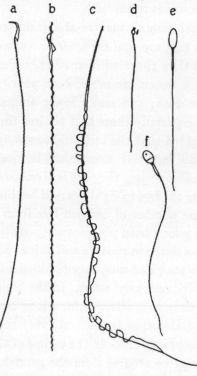

FIG. 6. Forms of spermatozoa. *a*, mouse; *b*, finch; *c*, triton; *d*, ascidian; *e*, horse; *f*, man. (After Ballowitz, Jensen, Broman.)

seemingly erratic fashion by the lashing action of its tail. Part of the irregularity in its movements is no doubt due to its coming into contact with other spermatozoa, or with small floating particles in the water. It has sometimes been described as swimming forward in a spiral path when free, and in circles when in contact with a solid.

If eggs are present in the water, the spermatozoa quickly accumulate in large numbers about the eggs, giving the impression that they are attracted toward the eggs. As a matter of fact, however, their accumulation is primarily due to their sticking to the membrane or jelly around the egg. Each spermatozoön that comes in contact with the membrane of the egg sticks to it; consequently in a very short time there may be hundreds around each egg. There is no direct evidence that the spermatozoa swimming near the egg turn toward it, and plenty of evidence that they may swim by the egg and pass on. Most of those in immediate contact with the egg generally take a vertical position with the head end in contact with the membrane of the egg. At first the tail continues to vibrate, but the spermatozoön that is destined to enter the egg becomes suddenly still, with its tail standing straight out from the egg in a radial direction.

When the spermatozoa are killed with suitable reagents and stained with special dyes, it is found that each is a rather complicated affair. In front of the head there is often a special body, called the acrosome, whose function is unknown, but may have something to do with the power of the spermatozoön to penetrate the membrane of the egg. The head is mainly composed of the nucleus of the cell from which the spermatozoön developed. This part consists of almost pure chromatin, the chromosomes being condensed into a tight mass. Behind the head is the middle piece, derived from the protoplasm of the original cell. It may contain one or more special substances, notably the centriole of the cell, or at least a body that becomes a centriole after the spermatozoön has entered the egg. The long tail can sometimes be seen to have an axial thread, bordered by a finlike membrane on each side. It is flattened in one plane. The motion of the sperm is due to the contractile nature of the tail. The spiral path that the sperm follows may be due to some asymmetry of the head or of the tail.

The entrance of the spermatozoön into the egg has often been watched, Fig. 7, but how it first succeeds in penetrating the surrounding membrane, often very tough, has never been discovered. After a few seconds the outer surface of the egg is seen to show a clear protoplasmic cone-shaped protrusion, exactly beneath the sperm head. In less than a minute as a rule the head passes through the membrane, often showing a constriction in its middle when half-

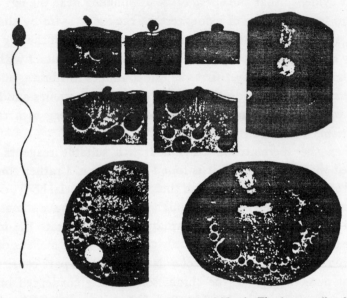

FIG. 7. Fertilization and polar-body formation of Nereis. The four smaller figures show entrance of sperm. The extrusion of the first polar body is shown in lower left-hand figure and of the second polar body in the two large right-hand figures. The last three also show the formation of the sperm asters, which is the beginning of the first cleavage spindle in the egg. (After F. R. Lillie.)

way through. Immediately on passing through the membrane its head is taken in by the fertilization cone, which in turn begins to retreat into the egg, carrying the sperm head with it. The tail in many eggs is left behind, sticking to the membrane, but in some eggs with soft membranes the tail also may be drawn into the egg. The middle piece, behind the head, goes in with the head, and may, as stated above, give rise to the centriole, which, after dividing, gives rise to the mitotic figure of the dividing egg.

In some eggs, as in the sea urchin, the spermatozoön may enter at any point of the surface; in others, as in the frog, it may enter only in the polar hemisphere; in others, as in ascidians, in the opposite hemisphere near the antipole. In some eggs with tough membranes

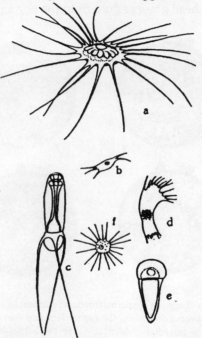

FIG. 8. Unusual forms of spermatozoa. *a*, crayfish; *b*, polyphemus; *c*, lobster; *d*, Sida; *e*, ascaris; *f*, moina. (From Korschelt and Heider.)

(fish eggs), there is a pore at one point in the membrane through which the spermatozoa must pass to reach the surface of the egg. In other eggs (insects) there may be several pores in the membrane.

While most spermatozoa, both in the lower and in the higher groups are threadlike, there are a few other kinds that have no tails. In the nematode worm, the ascaris of the horse, the spermatozoön is said to be amoeboid, Fig. 8e. Fertilization is internal, the semen of

the male being transferred to the oviduct of the female. In some
of the higher crustaceans the spermatozoa have varied shapes, Figs.
8a–f, and there are various devices by which the semen is brought
into contact with the surface of the egg.

The passage of the sperm nucleus from the periphery toward the
center of the egg takes place in very much the same way in all animals. The sperm head absorbs fluid from the egg and slowly enlarges

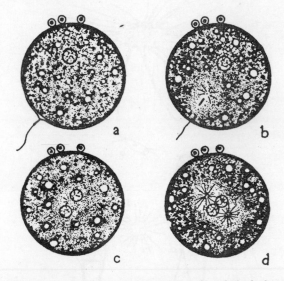

FIG. 9. Fertilization of egg. In *a*, spermatozoön at surface; in *b*, the head has entered
and a sperm aster appears near it; in *c*, the egg nucleus and sperm nuclei are approach-
ing each other and the aster has divided; in *d*, the two nuclei have met, and the two
poles of mitotic spindle are seen.

until, in most cases, it is as large as the egg nucleus, Figs. 9c–d,
which has also moved from the pole toward the center of the egg,
where the two pronuclei meet, and generally unite to form a round
nucleus.

THE GERM TRACK

The germ cells, eggs and spermatozoa, can generally be traced
back to an early stage in the development of the embryo. On this

fact Weismann based his well-known theory of the isolation of the germ cells, which are early set aside for the continuity of the race, while the other cells of the embryo are differentiated into the soma or body of the embryo. The germ cells are immortal, the somatic cells have only a limited length of life. It is through the line of germ cells that the organisms of today are connected with the oldest forms of life, whose bodies have perished. However, since all the heritable characteristics of the race are carried in the chromosomes of all of the cells, both germ cells and body cells, this distinction no longer

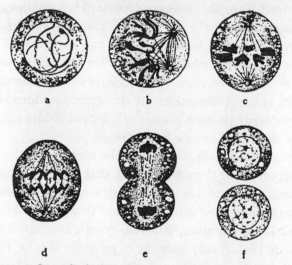

FIG. 10. Stages in the karyokinetic or nitotic division of a cell.

has the importance ascribed to it by Weismann. It is true, nevertheless, that the germ cells, as such, are set aside as early or even earlier than the other cells that become differentiated into the somatic organs of the embryo.

At first the germ cells do not increase rapidly in number, but in the later stages of the embryo or in the adult—the time differing greatly in different types—the germ cells divide and increase in number, and this may continue throughout life, especially in the male. These early divisions of the germ cells—called the oogonia and spermatogonia—take place by the ordinary process of division, Fig. 10. Each contains at this time the double set of chromosomes, and at each division each chromosome splits lengthwise into equal halves, one daughter chromosome passing to one cell, one to the other. When they have passed through a number of ordinary divisions, they then undergo two unique divisions in one of which the number of chromosomes is reduced to the half number. These divisions are spoken of as the maturation divisions, or technically as meiosis, in contrast to mitosis or the ordinary process of cell division. The two maturation divisions of the egg cell take place after it has accumulated more or less yolk material. These two divisions of the egg have already been described. They involve the giving off of two polar bodies. The two divisions of the sperm cells are essentially the same, so far as the chromosomes are concerned, but four functional cells are produced from each mother cell, instead of only one as in the egg. These divisions may now be described.

Just before maturation division of the sperm cells the chromosomes reappear as fine threads, Fig. 11a, that extend throughout the large nucleus. Some of the chromosomes are loops, others rods. They come to lie side by side throughout their length, Figs. 11b–c, and appear to fuse together, but from genetic evidence it is known that they do not actually fuse. It is probable that in this stage, either in the eggs or in the sperm cells or in both, an interchange of blocks or pieces of exactly equivalent parts takes place. The process is called crossing over, Figs. 12a–d. As a result, a recombination of the genes in homologous chromosomes is brought about. The thread then shortens and the nuclear wall disappears. A spindle appears in the protoplasm, and the chromosomes become attached to its fibers, Fig. 11d. There seem still to be only half as many chromosomes present that are characteristic of the species, and at this time they can often be seen to be split throughout their length; in fact sometimes a secondary split is observable giving what is called a tetrad.

The cell then divides, Figs. 11e–f–g. At this division the pairs that are united may separate, but it appears that in some cases the separation may be along the secondary split, or in some chromosomes the separation is one way, in other chromosomes the other way. A short resting stage follows, and then each cell divides again, Figs. 11h–i. As a result of these two divisions four cells are formed each containing a set of chromosomes, *i.e.*, one of each kind. Then

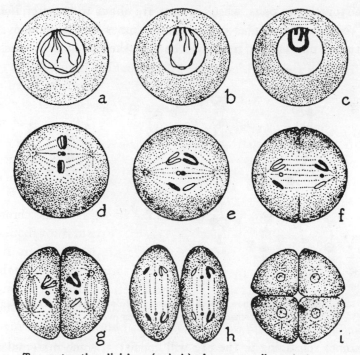

FIG. 11.   Two maturation divisions (meiosis) of a sperm cell, producing four sperm cells.

follows an elaborate process of differentiation by which each cell is changed into a functional spermatozoön.

It is evident that the maturation divisions in the egg and sperm cells are essentially the same as far as reduction in number of the chromosomes is involved. It has been shown, both by observation and by deduction from genetic evidence, that the separation of the conjugated members of each pair of chromosomes is at random. A redistribution of the inherited maternal and paternal chromosome takes place at the maturation division, both in eggs and sperm cells.

The importance of these facts for genetic theory is most easily shown in cases where the individual is a hybrid—that is, one that has received from its mother one haploid set of chromosomes, and from its father another different haploid set. Its chromosomes come together in pairs, when the eggs are about to mature. If now the reduction in number of the chromosomes (described above) is a random one, the end result will be that the half number of chro-

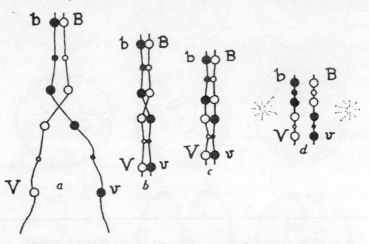

FIG. 12.   Diagram to show how two chromosomes coming together may overlap and interchange segments (crossing over).

mosomes remaining in the egg will consist of some maternal and some paternal chromosomes; for seldom will it happen, by chance, that they are all maternal, or all paternal.

Thus, in every generation, the ripe eggs of a hybrid individual will contain all possible combinations of the chromosomes, one of each kind, that were derived from the mother and father of the individual. The offspring, taken as a whole, will be made up of all possible combinations of their grandparents' chromosomes. This process of sorting out the chromosomes and their reunion at fertilization furnishes the mechanism of Mendel's two laws of heredity.

CHAPTER IV

CLEAVAGE OF THE EGG

About an hour after fertilization the egg begins to divide. The divisions of the egg are called cleavages. The egg is also said to segment and the early formed cells are often called blastomeres. Most eggs divide into two equal parts, but the first division of others is into two unequal parts. Both types are typical cell divisions with respect to chromosomes, spindles, etc., but while the successive divisions of the cells of the adult take place only after a long resting period, during which the daughter cells grow to the original size of the mother cell, the interval between divisions of the egg is very short, and the daughter cells do not grow larger before the next division.

The second division of the egg follows the first in about an hour, or somewhat less, and this rate is approximately kept until a large number of cells, often several hundred, or even a thousand or more, are produced without the total mass of protoplasm becoming any bigger than it was at the beginning. This is one of the most distinctive features of the cleavage of all eggs. The second division is at right angles to the first and also through the pole, sometimes giving four cells of equal size, but not infrequently two of them may be a little smaller than the other two. The third division is at right angles to the first two. It lies at or near the equator in typical cases. Thus the first three divisions are in the three dimensions of space.

The speed with which the divisions follow each other is, within certain narrow limits, a function of temperature. If the eggs are kept cool the division rate is slowed down, or even entirely stopped; if the eggs are kept at a temperature a little above that of their normal environment they divide more rapidly, but the upper temperature limit is quickly reached, because apparently some of the important constituents of the egg are coagulated. In fact, the eggs of most species of animals are "adapted" rather closely to the average temperature of the medium in which they develop.

The medium is an important factor in the development of the egg. Between the egg and its environment there are constant interchanges, of which the most important is respiration, taking in oxygen and setting free carbon dioxide. This interchange is regulated by the surface of the egg itself, and also by the egg membranes. These, and other surface relations, play an important rôle in development. The salt constituents of the egg and those of the environment may be quite different, and the difference is also regulated largely by the surface membrane of the egg itself. In marine eggs especially the salts of the sea play an important part in development. The eggs of insects are often laid in dry places, and the eggs themselves contain all the necessary constituents for early development.

THE EGG OF THE SEA URCHIN

In order to gain further insight into the kind of processes that take place in the egg at the time of cleavage, it is necessary to examine some of the visible movements that go on in it, before and immediately after fertilization, even before the first cleavage takes place. The egg of the sea urchin, Paracentrotus, illustrates very well some of these changes. When the egg is still attached to the wall of the ovary, its outermost layer contains numerous small reddish granules, Figs. 13a, b, evenly distributed over its surface. When the egg is set free in the cavity of the ovary and the nuclear wall breaks down, preparatory to the extrusion of the polar bodies, the granules of the upper hemisphere are carried below the equator of the egg to form a ring, Fig. 13c. This means that the surface protoplasm in which the granules lie moves over the egg from the upper hemisphere into the lower. The ring remains in this position until fertilization takes place. As shown in Figs. 14a–h, it still holds this position during the following divisions of the egg, and finally, when gastrulation takes place, the red cells are turned into the interior of the egg and become the walls of the digestive tract, Figs. 14l–n. Evidently considerable rearrangement of the constituents of the egg takes place before cleavage, and there is evidence in other eggs showing that similar movements are going on during and after fertilization.

The first three cleavages bear a constant relation to the polar axis of the egg, an imaginary axis passing from the pole (where the polar bodies are set free) through the center of the egg to the opposite side, the antipole. The stratification of the materials of the egg is at right angles to this axis. The first division of the sea urchin's egg passes through the pole, Fig. 14a, the center of the egg, and the antipole. The second division, at right angles to the first, also passes

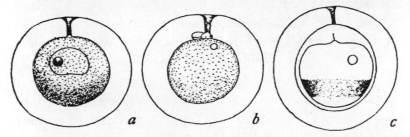

FIG. 13.   Movement of the red pigment of the egg of Paracentrotus at the time of fertilization. (After Boveri.)

through the pole, Fig. 14b. The third division is near the equator of the egg, Fig. 14c. The eight resulting cells are approximately equal.

The location of these planes of division is indicated before each division actually appears by the position taken by spindles, which lie in planes at right angles to the forthcoming divisions. This raises the question as to whether the spindle determines the plane of division, wherever it may be, or whether the latter, being fixed, the spindle moves into the proper position before divisions of the egg come in. Fortunately there is evidence bearing on this question that comes from compressing the egg and from centrifuging it. This evidence, if not entirely convincing, answers the question to some extent.

For example:—By slightly compressing the egg between the glass slide and the cover-slip, the position of the cleavage planes can be altered; the first, second, third, and even later cleavages, Figs. 15a–e, come in at right angles to the compressing plates. Should a flattened egg happen to lie with its pole against the cover slip, the first two

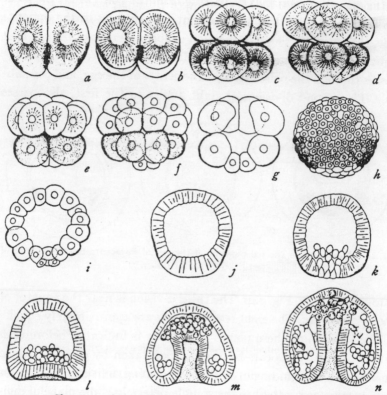

FIG. 14. Cleavage and gastrulation of the egg of Paracentrotus. (After Boveri.)

divisions are through the pole, and *may* happen to lie in the same meridian in which they would have lain if not compressed. Under these conditions the first spindles would be in their usual positions. But if an egg happened to lie on its side, i.e., with the pole 45° from the compressing plates, the situation is different; the first division

still passes through the pole and at right angles to the plates, and
since it also passes through the equator of the spindle, the spindle

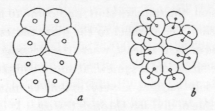

FIG. 15.   Egg of sea urchin under continuous pressure kept alive by a stream of sea
water. (After Ziegler.)

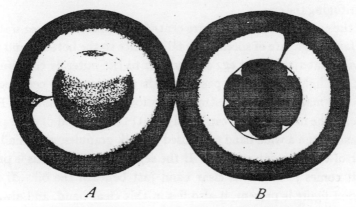

FIG. 16.   Centrifuged eggs of the sea urchin, Arbacia. *A*, stratified egg of Arbacia surrounded by the clear jelly; the funnel lies at one side. *B*, an eight-cell stage with micromeres opposite funnel. (After Morgan and Spooner.)

must in many cases have shifted, or developed in a new position.
This is more evident in the case of the second division of the compressed egg when the cleavage plane does not pass through the pole

at all, but lies at right angles to it, i.e., at the original equator of the egg where the third normal cleavage plane lies. It follows that, under compression, the spindles shift, and come to lie at right angles to the compression, i.e., parallel to the glass plates. Their location, then determines in part, at least, where the cleavages will take place.

Of course the situation is somewhat more complex than this, for the spindle is only part of a very extensive mitotic figure that extends at the time of division through the egg. It is known that the whole structure, the mitotic figure, may be shifted under compression and carries the spindle proper with it into the new position. It is the whole division figure, not the spindle alone or its attached chromosomes, that is responsible for the result. It is known that the division figure is a gel that is firmer than the rest of the material of the egg in which it floats, so to speak. It can be mechanically shifted about as a whole, and probably can to some extent also be reformed in a new position. The most convincing evidence of this comes from centrifuging the egg.

If the eggs of the sea urchin are centrifuged in a tube of sea water, at a fairly high rate of speed, they fall at once to the bottom, and the heavier materials of the egg slowly pass to the outer or centrifugal end; the lighter materials pass towards the central or centripetal end. The materials become stratified in four layers as shown in Fig. 16A. At the inner end is a cap of oil or fat; then a band of clear protoplasm; then a wide band composed of yolk granules, and finally a layer of red pigment granules. If the segmentation nucleus is present it comes to lie in the clear band just beneath the oil. If the division figure is present, it also lies in this clear band, and always parallel to the stratification. When cleavage takes place, Fig. 17a, it cuts at right angles through the stratified layers. There is a funnel in the jelly around the egg, that can be demonstrated by mixing sepia ink with the water. The funnel locates the pole of the egg, Fig. 16A, and in this way it can be shown that the first cleavage plane of the centrifuged egg bears no fixed relation to the pole. It follows that the location of the cleavage plane is determined by the position of the spindle. In other words: in the normal egg it is the division figure that adjusts itself to the polar configuration, hence the first division passes through the pole. The plane of division, as such, is not predetermined by the pole, or at least it may be shifted into a new position.

The second division of the centrifuged egg is at right angles to the first, and parallel to the stratification, Fig. 17b, which means that the spindles assume a position at right angles to the first one. Two of

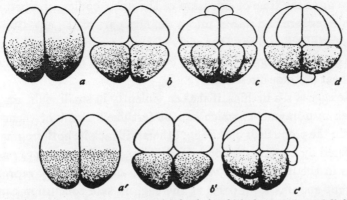

FIG. 17.   Cleavage of centrifuged eggs of Arbacia in which the first plane of division is at right angles to the stratification, and the second at right angles to the first and parallel to the stratification. The third cleavage is in the plane of the paper. The fourth cleavage is shown in *c, d* and *c'*; the micromeres may lie at the centripetal pole (*c*), at the centrifugal pole (*d*) or at the side (*c'*). (After Morgan.)

the cells will contain all of the oil and most of the clear substance; the other two cells will contain the yolk and pigment. The third division is at right angles to the other two, giving four cells of one kind and four of the other kind.

The next division is significant. The four micromeres appear at a crossing point of two of the division planes, Figs. 17c–d. They may lie where the first and second, or the first and third planes cross; they may lie in the pigment field, or in the yolk field, or between the two. Whatever their position may be, it has been found that the micromeres are formed opposite, or as nearly so as possible, to the funnel. This seems to mean that, at the time of their appearance, there is a region of the egg that gives rise to the micromeres, and that their formation at the fourth cleavage is largely independent of the sequence in which the preceding planes have appeared.

The micromeres at a later stage wander into the interior of the blastula Figs. 14k–n, and become the mesenchyme. It may appear, therefore, that there is a definite region of the egg destined to become micromeres, but whether it exists localized at the antipole before or at the time of extrusion of the polar bodies, or whether it arises elsewhere and moves into a definite part of the egg after the egg is fertilized, or during the early cleavage period, is not shown by these centrifuged eggs. There are, however, some other experiments that bear on these questions.

The eggs of sea urchins, if shaken violently in small vials, may be broken into fragments which can be fertilized. Some of them will contain the pronucleus of the egg; others will lack it; both fragments, if entered by a spermatozoön, will develop. The formation of micromeres in the fragments is very irregular, and since they represent different and unknown parts of the egg, no safe conclusion can be drawn from such results. A more precise method consists in cutting the egg into two pieces, and following the history of each piece. Unfortunately, in most sea urchin eggs, the pronucleus does not, before fertilization, lie at any fixed point with respect to the pole; the polar bodies are lost, and the funnel can not be seen in the jelly without treatment with sepia ink. In one species, however, Lytechinus, the polar bodies are present, which makes orientation of the cut possible. When these eggs are cut in a vertical plane, i.e., through the egg's axis, both pieces develop after fertilization; one is haploid, the other diploid, but except for their size the embryos (plutei) can not be distinguished from normal ones. When both fragments divide, the pattern of one of them is seen to be like the normal, the other has no micromeres. This seems to mean that the cut was to one side of the axis, and the antipole lay in one fragment only; or else it may mean that the micromere material was at this time in or near the center of the egg, and lay to one side of the cut. Other and later experiments make a decision difficult, since the authors concluded that there is no localized micromere-forming material in this egg—at least, no such material has been "differentiated" before fertilization. That such material, may, nevertheless, be present in the egg, even if not diffused over the surface, or concentrated near the center of the egg, has not been convincingly shown. There are other experiments on the egg of the sea urchin, Arbacia, that seem to show that such material is present in the interior, even before fertilization takes place. In this connection it should be recalled that, in the egg of Arbacia, the material out of which the micromeres form wells up from the interior of the egg at the antipole when the spindles for the micromeres approach this region. In fact the micromeres consist almost entirely of the outer polar aster; only the thin surface layer is made up of the original polar material.

THE ASCIDIAN EGG

Since the early observations of Edward Van Beneden on the eggs of an ascidian they have been a favorite object for experimental work. Some of these eggs are as transparent as glass; others contain pigment granules whose movements have furnished excellent landmarks, and the very early differentiation of their cells makes it possible to follow the cells to their destinations. There is one drawback, namely, a tough membrane about the egg. There is, in addition, an inner circle of cells, Fig. 18a, the test cells, between the surface of the egg and the membrane, and another circle of cells over the outer surface of the membrane. A method has recently been found to digest off the membrane without injuring the egg.

The ripe eggs leave the ovary and collect in the oviduct. The polar spindle is present at the time, but the polar bodies are not given off until a spermatozoön enters the egg. It enters at or very

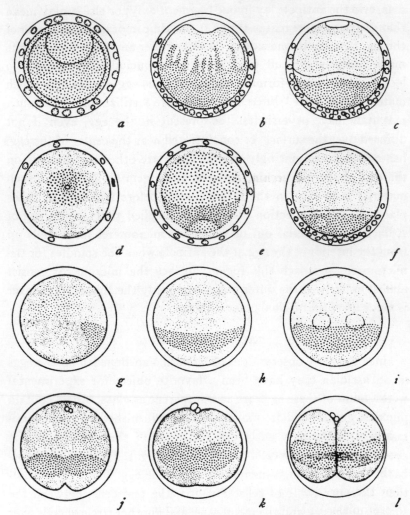

FIG. 18.   Movements of the materials of the egg of the ascidian, Styela, at the time of
fertilization. (After Conklin.)

near the antipole. The egg of Styela contains yellow pigment granules in its surface layer, Fig. 18a. When the spermatozoön enters the
egg, the outer layer of protoplasm moves rapidly over the surface

carrying with it the colored granules, Fig. 18b. A yellow crescent is formed on one side, extending more than halfway around the egg, Figs. 18h–l. The crescent is the first indication of a bilateral appearance in the egg.

The first division plane passes through the middle of the crescent, Fig. 18l, half of the crescent lying in one cell, half in the other. Whether the surface protoplasmic layer and the yellow pigment of the crescent pass to a predetermined side of the egg, or whether their location is determined by the entrance point of the spermatozoön is still a matter of dispute. One point, however, seems fairly well established. The head of the spermatozoön first moves toward that side where the middle of the crescent comes to lie. If, as seems to be the rule, the spermatozoön does not enter exactly at the antipole but a little to one or the other side of this point, the excentric point of entrance may be the determining factor that leads to a more exten-sive accumulation of superficial protoplasm (and pigment) on the same or on the opposite side. If this reasoning is correct, the bilateral arrangement of the contents of the egg is not predetermined, but is secondarily acquired, as in some other eggs, by the point of entrance of the spermatozoön.

The first division of the egg passes through the pole and through the middle of the crescent, Figs. 19a–b, dividing the egg into two exactly equal parts. The second division is also through the pole, and at right angles to the first plane of division. Two of the cells are somewhat larger than the other two and contain the greater part of the crescent, Fig. 19c. The two smaller cells contain only the horns of the crescent, and may be spoken of as the anterior cells. The protoplasm of the anterior and posterior cells also differs somewhat in appearance, and, no doubt, their contents differ in other respects than in the visible amount of pigment, or yolk, contained in them.

The third division is at right angles to the first two, and cuts off from each cell, Fig. 19d, a smaller cell at the polar end, and a larger cell on the opposite end of the egg. Two of the latter contain most of the yellow band of the crescent. From this time onward the divisions continue to take place in an orderly way, producing cells that can be identified by their contents and location. The fate of these cells will be described in the chapter on gastrulation.

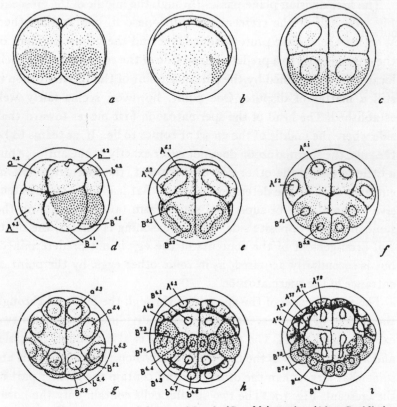

FIG. 19. Cleavage stages of the egg of Styela (Cynthia) partita. (After Conklin.)

The few compression experiments that have been made on the ascidian egg, show that these eggs, as do others, divide at right angles to the compressing plates, but unlike the compressed eggs of the sea urchin, none of the eggs, when freed from pressure, produce wholly normal embryos, even although they continue to divide, and some of them give embryos with dislocated organs. The simplest explanation is that in these eggs that differentiate very early, any disturbance of the orderly sorting out of the materials leads to disaster.

The effect of centrifuging these eggs has also been studied. It is more difficult to bring about stratification than in the case of many other eggs, because of the greater viscosity of the protoplasm. If the eggs of Styela are centrifuged before fertilization, the yellow pigment may be driven to one pole; presumably the superficial protoplasm undergoes its normal movements. These eggs may produce normal embryos. The pigment may come to lie in any part of the embryo. This proves that the yellow pigment, as such, is not organ-forming. Nevertheless, in the normal egg its movements indicate changes in the redistribution of the protoplasmic contents of the egg that are correlated with subsequent effects on development. In other ascidians, lacking pigment, such as Ciona, there is evidence to show that movements of the superficial protoplasm, leading to the formation of a crescent, take place as in Styela.

It is difficult to obtain fragments of the egg of ascidians from recognizable regions because of the tough membrane, and until recently it has been impossible to cut fragments from the egg from known locations. Nevertheless, it has been shown that the fragments will divide and it is claimed will form normal embryos, but until the relation of the fragments to the regions of the egg has been accurately determined, it is unsafe to generalize from these inadequate results.

CLEAVAGE IN OTHER TYPES

Other types of cleavage than those of the sea urchin and the ascidian are well known. The spiral type of annelids and molluscs differs from the preceding ones mainly in that, at the third cleavage, four cells, the micromeres, are given off around the pole in a right or in a left-handed spiral, Fig. 20. At the sixth cleavage a special asymmetric cell, called d⁴, is laid down at the equator that gives rise to the mesoderm, or middle layer, of the embryo, Fig. 20. Its later division into two marks the first strictly recognizable bilateral plane of the embryo.

The females of the marine annelid worm, Nereis, come to the surface on certain nights of the summer months and swim about

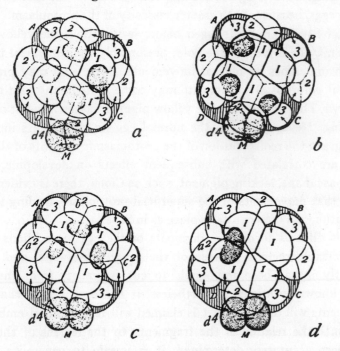

FIG. 20. Late stages of cleavage of egg of planarian, *a*; annelid *b* and two molluscs *c* and *d*. The position of the *d⁴* cell is shown. (After Wilson.)

very rapidly. The males also come to the surface at the same time. If one approaches or strikes the female she immediately sets free all of her eggs and the male pours out his semen at the same time. The eggs are fertilized in the sea water. Each egg is transparent, and when set free it contains a large egg nucleus which quickly disappears on fertilization, setting free its chromosomes. T.

surrounded by a thick membrane with radiating pores. Under the membrane there is a layer containing a fluid substance in vacuoles, which passes through the membrane at once when the spermatozoön touches the surface of the egg. The fluid coagulates in sea water, and becomes the gelatinous covering of the egg. It continues to swell for

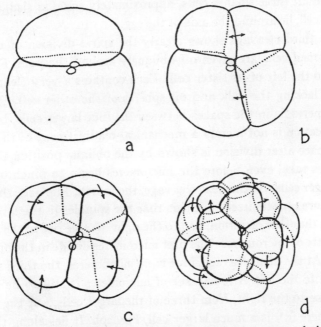

FIG. 21. Cleavage of egg of Nereis. The formation of the first quartette of micromeres is seen in *c*, and the second quartette in *d*. In the latter the first micromeres are also dividing. (After Wilson.)

several hours, until the diameter of the egg plus jelly is at least three times as great as that of the original egg.

It takes the spermatozoön over half an hour to penetrate the membrane of the egg. Its head is taken in by a conspicuous fertilization cone; the tail is left outside and serves as a landmark by which the position of the first cleavage plane in relation to he point of entrance of the sperm can be determined. The two coincide in a high percentage of cases, and approximately so in most of the remaining ones. Two polar bodies are given off inside the membrane. The first cleavage, Fig. 21a, divides the egg into somewhat unequal parts. The second cleavage, at right angles to the first, divides the smaller cell into equal parts and the larger cell into unequal parts, Fig. 21b. As a result three smaller cells, approximately equal in size, and one larger cell, lie around the axis of the egg.

The third cleavage shows clearly the spiral division of the egg. Four small cells are given off obliquely around the pole, Fig. 21c, each to the left of its sister cell. Each contains a very clear protoplasm lacking the yolk and oil spheres of the other cell. The four micromeres lie in the spaces between the four larger cells, but that this location is not merely a mechanical readjustment of the available space after division is shown by the oblique position that the spindles take, even before the micromeres begin to pinch off from the larger cells. At the next cleavage, the fourth, Fig. 21d, the same procedure repeats itself, but this time the spindles in the large cells turn in the other direction, i.e., to the right, to give rise to the second quartette of micromeres. The first micromeres divide in the opposite spiral. At the fifth cleavage, four more micromeres, the third set, are formed to the left. A fourth set of micromeres is still to be added, this time to the right, from three of the larger cells, but the fourth micromere (d^4) is a much larger cell, Fig. 20b. It lies along the line of the original second division. This cell is the forerunner of most of the mesodermic cells of the embryo. When it divides into two equal cells, the middle line of the embryo, lying between them, is established. This middle line does not correspond exactly with either of the first two divisions, but is very near the plane of the second division. The micromeres continue to divide and finally cover the surface of the embryo. The large yolk-bearing cells give rise to the archenteron.

The division of the eggs of most of the molluscs, Figs. 20c, d, except those of the squid and octopus that have large eggs, is similar in

principle, often even in great detail, to the divisions of the egg of
Nereis, despite the fact that the adult animals belonging to these
two great groups are so entirely different. Yet the similarity of the
earliest stages and the fact that the origin of the organs is so much
alike can leave little doubt that the two groups are related, and in a
far past, when there is no record of a common group of fossil forms,
their common ancestors passed through stages of development similar to those shown by both types living today.

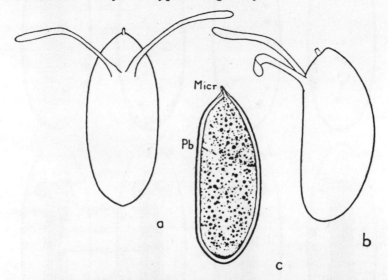

FIG. 22.   *a*, Bilateral egg of Drosophila, as seen from dorsal side; *b*, from right side; *c*,
section of egg in median plane and through the micropyle (*micr*). The polar bodies have
been given off on upper third of dorsal side.

The eggs of insects develop in an entirely different way. If the
insects have come from annelid ancestors, as is commonly taught,
they have acquired a very different type of cleavage. The eggs of
most insects are not spherical, as are the eggs of most marine and
fresh-water animals, but are elongated, taking their shape, in the
later stages of their growth, from the egg tubes in which they are
formed, Figs. 22a, b. This shape is maintained by the very hard coat

surrounding them. The coat is pierced by one or more pores which lie, not at the end of the egg, but on one of its sides. Many of the eggs are also bilateral, one face being flatter than the other, Figs. 22b, c. These faces correspond to the ventral and dorsal sides of the

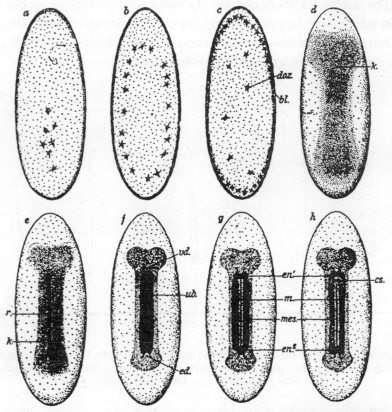

FIG. 23. Cleavage and formation of the embryo of an insect. (After Mangold.)

embryo. The bilateral form of the egg is present when the egg is ready to be laid. As the egg enters the posterior part of the oviduct it passes the opening of the seminal receptacle with the micropyle next to the opening. A few spermatozoa enter the egg, both head and tail. The egg gives off its polar bodies and the egg nucleus sinks deeper into the protoplasm. The sperm nuclei move toward the egg nucleus, and one of them comes into apposition with the latter. The other sperm nuclei then stop their movements toward the egg nucleus, and later degenerate.

The interior of the egg is filled with yolk granules imbedded in the protoplasm. The surface of the egg is covered with a layer of protoplasm that is free from yolk, Fig. 23a. A spindle develops near the conjugate nuclei, and a small amount of protoplasm collects around

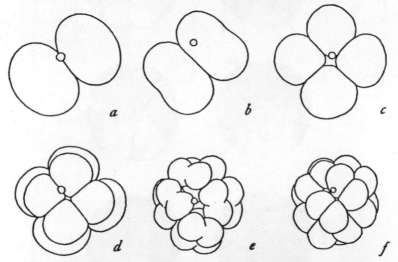

FIG. 24. First four cleavages of the egg of Amphioxus. (After Cerfontaine.)

them. The nuclear walls disappear, and the chromosomes pass on to the spindle where they divide and pass to the poles. Two nuclei result, but the protoplasm does not divide. These nuclei divide again, and the process continues until a number have developed, Figs. 23a, b. They move apart after each division, and soon come to form a ring, or rather a sphere, Fig. 23b. As their number increases they move nearer the surface of the egg, and come finally to lie in the superficial layer of protoplasm, Fig. 23c. At this time several hundred nuclei are present. Then the superficial protoplasm begins to cleave, each furrow sinking in between the halves of a dividing nucleus. These furrows pass in radially, but stop at the inner side of the protoplasm, each cell remaining open at its inner end, its contents being continuous for some time with the central yolk mass. At

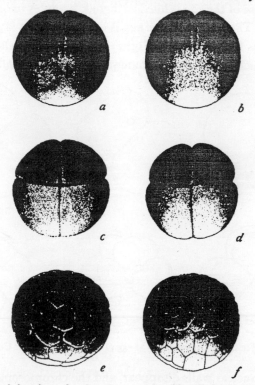

FIG. 25.   Pairs of drawings showing the opposite sides of the frog's egg during the early cleavages. The darker side, in the left-hand figures is the so-called ventral side of the egg, and the lighter side, in the right-hand figures, is the so-called dorsal side or the side of the gray crescent. In *e*, and *f*, the cells on the latter side are smaller than those of the opposite side at the same level. (After Brachet.)

the posterior end of the egg, one or a few nuclei pass into a special protoplasm; these cells are the forerunners of the germ cells.

The eggs of the lower crustacea are encased in hard shells. They divide into two, four, eight cells etc., but those of the crayfish and the lobster show much the same type of cleavage as that of the insects. There is a central yolk mass and a superficial layer of cells. The cell walls between the latter extend to the center of the egg.

In the vertebrates there are two principal types of cleavage. Amphioxus, standing at the bottom of the group, has small eggs that divide into two equal parts, then four, eight, etc. Figs. 24a–d, much as in the ascidians. The frog's egg, which is much larger, has also a total cleavage, Figs. 25a–f. There is one special feature of its early development that recalls that of the ascidian egg. As soon as the spermatozoön enters the egg in the upper hemisphere, a move-

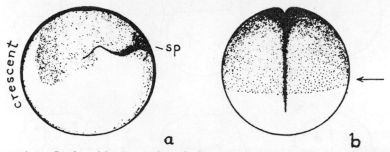

FIG. 26. *a,* Section of frog's egg through the crescent in plane of first cleavage. The penetration path of the spermatozoön is seen on the side of the egg opposite the crescent; *b,* surface view of frog's egg dividing into two. The cleavage cuts through the crescent. (After Schultze.)

ment of the superficial material takes place on the side of the egg that is opposite the entrance point. A gray crescent appears along the border between the black and the white hemispheres, Figs. 26a, b. The middle of the crescent, as in the ascidian, becomes the median plane of the embryo. Through its middle line the first division takes place, as a rule; but whether this division does, or does not, pass through the middle of the crescent—it may rarely be 90° away—the mid-plane of the embryo corresponds to the middle of the crescent.

Most of the eggs of teleostean fishes are smaller than the frog's eggs, but nevertheless do not have total cleavage. The egg is made up of a central yolk mass, covered by a rather thin layer or cap of

protoplasm which is slightly thicker beneath the pole. There is a
membrane over the egg pierced at the pole by a small tube through
which a spermatozoön enters. The first cleavage, Fig. 27b, consists
of a straight furrow in the polar cap of protoplasm; the second follows at right angles to the first; the third is at right angles to the
second, Figs. 27c, d. The following divisions are somewhat irregular
in position, Figs. 27e, f. The first furrows do not pass at first very

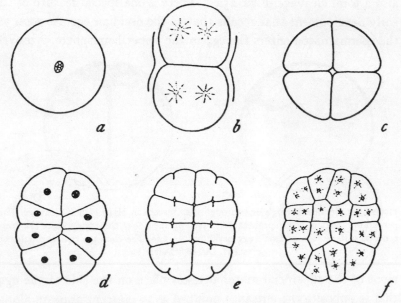

FIG. 27.   The first four cleavages of the egg of a teleost fish. (After H. V. Wilson.)

deep into the egg; after a few superficial furrows have been formed
the more central cells divide at right angles to the last, forming an
outer cell and leaving an inner nucleus in the superficial yolk.

The eggs of reptiles and birds are relatively enormous. The yolk is
the egg proper, around which a jelly (the "white") is laid as the egg
passes down the oviduct. Over these the calcareous shell is deposited still farther down the oviduct. Fertilization takes place as
soon as the egg leaves the ovary and enters the oviduct. The cleav-

age consists at first of superficial furrows as in fish eggs. A disc of many cells results, lying at the pole of the egg.

The mammals have descended from forms having large eggs like those of reptiles. The lowest living type of mammal, Platypus, still retains this type of egg, measuring three millimeters in diameter, but all the others have almost no yolk, and the eggs, in consequence, are very small, barely visible to the naked eye. The mammalian egg

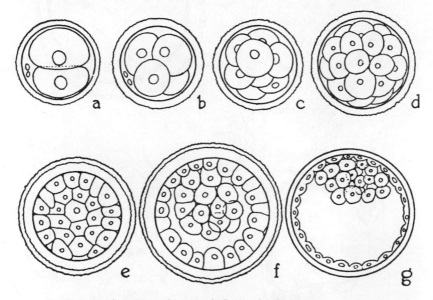

FIG. 28. Segmentation of egg of mammal. In f and g the formation of the blastocyst and blastoderm has begun. (After Prentiss.)

divides totally into two, four, eight cells nearly equal in size, Figs. 28a–c. A fluid appears inside the mass, Fig. 28f, and the cells come to cover the walls of a vesicle, Fig. 28g, that gradually increases to a considerable size, Figs. 29a, b. At one point some of the early formed cells slip under the outer layer, which then comes to cover them, Fig. 29a. This inner group of cells then forms a small disc which later gives rise to the embryo, Fig. 29d. The walls of the vesicle become attached to the walls of the uterus of the mother, Fig. 29c, where the embryo undergoes its later development.

There are two stages in mammalian development that would seem

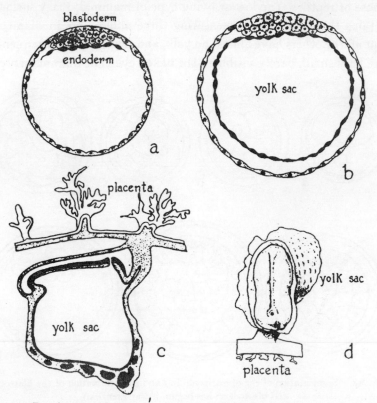

FIG. 29. Development of mammal. *a*, Blastocyst with blastoderm above; *b*, later stage with endoderm forming "yolk" sac; *c*, embryo of later stage with "yolk" sac and placenta; *d*, surface view of early embryo. (From Keibel and Mall.)

meaningless without a knowledge of the descent of the mammals from forms having large eggs.

The very young cells that will become eggs lie at first in the superficial layer of the ovary, Fig. 30a. Some of them sink into the interior of the ovary, carrying with them a column of other cells that come to surround each egg. As the egg cell grows larger a fluid space appears between the cells immediately around the egg and the outer layer of cells, Figs. 30c, d, e; the space becomes relatively enormous, and the sphere, called a Graafian follicle, protrudes from the surface of the ovary. Except for the fluid in the follicle, it resembles the large follicle of the ovary of the bird and lizard in which the yolk-bearing egg entirely fills the interior of the follicle. If we interpret

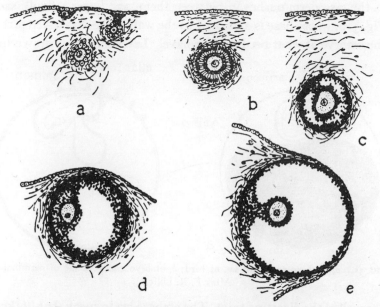

FIG. 30.   Four stages of egg and Graafian follicle of mammal. (From Patten.)

the size of the follicle as a reminiscence of the ancestral condition, when the mammalian egg contained yolk, we get a sort of explanation of the present-day conditions. This explanation postulates that the mammalian egg retains today some of the stages of development of its ancestors.

When the egg is ripe, the follicle bursts. The funnel-shaped mouth of the oviduct, opening into the body cavity, grasps the ovary and receives the egg, which is carried along the anterior portion of the tube, and later fixes itself to the uterine wall, where its later development takes place.

The other peculiar stage of development referred to concerns the development of a yolk sac that is yolkless. When the embryo is about to appear on the blastoderm, Fig. 29b, the endoderm edges have extended around the inner walls of the ectodermal vesicle to form an inner sphere, or sac; this shrinks away from the outer wall as the mesoderm pushes in between them, and forms a "yolk sac," Fig. 31b, whose base is attached to the wall of the digestive tract of the embryo at what becomes the navel. Later this sac is drawn into

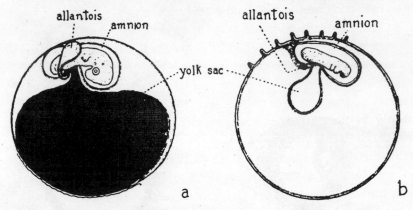

FIG. 31. a, Embryo and yolk sac of bird; b, embryo and yolk sac of mammal. (a, After F. R. Lillie.)

the wall of the digestive tract. The embryo, so to speak, has "faked" a yolk sac, although the sac has no yolk. In other words, the mammal develops as though it had come from a large yolk-bearing egg like that of the bird, Fig. 31a, or reptile. It follows still the ancestral method of development. The yolk sac is a rudimentary organ, or an embryonic ancestral reminiscence. In other words, the gene-complex involved in this stage of development has not been changed. It would certainly appear simpler, then, to assume that special genes are concerned with these stages whose action has not been affected by changes in other genes that brought about the later evolutionary stages.

CHAPTER V

GASTRULATION

The process called gastrulation involves the turning in of the outer wall of the hollow sphere or blastula to make the digestive tract. Gastrulation is seen in its simplest form in the development of the sea urchin, Figs. 14l–n, or of amphioxus, Figs. 33a–c, and in its most modified forms in birds and mammals.

At the end of the period of most rapid cell division of the egg of the sea urchin, a hollow sphere (the blastula), Fig. 14i-j, has been formed whose wall consists of a single layer of short columnar cells. The cells are smaller in the polar, and larger in the antipolar hemisphere. In Paracentrotus the pigment ring lies near the equator, but mostly in the antipolar hemisphere. A group of micromere cells wanders into the interior of the blastula to produce the mesenchyme cells, Figs. 14k–n, from some of which the larval skeleton soon develops. The cells of the lower hemisphere next turn in, i.e., invaginate into the interior as a tube, called the archenteron, which extends almost to the opposite wall of the gastrula.

From the innermost end of the tube an evagination is pinched off that constricts into two sacs. These form the body cavity and water vascular system of the larva. The inner end of the archenteron next bends over to one side and comes into contact with the inner wall of the blastula. At the point of contact the blastula wall turns inward in the form of a small pouch or short tube, which becomes the mouth and oesophagus of the larva.

The blastula stage is to all appearances at first radially symmetrical around the polar axis. There are no visible indications of a future bilateral organization, but there is some experimental evidence to show that in a large percentage of cases it is present, the plane of symmetry corresponding to the plane of the first cleavage. There are several indications that this relation holds, but the most convincing evidence has been obtained by means of intra-vitam staining. Eggs in the two-cell stage are brought in contact with small plates of agar-agar that have previously been soaked in Nile blue, Figs. 32a–b. The part of the egg in contact with the plate becomes blue. When

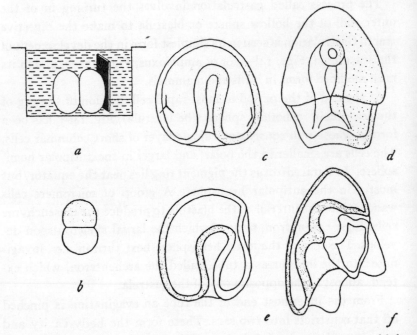

FIG. 32. Method of staining the egg of a sea urchin in the two-cell stage, *a* and *b*, by means of a block of agar colored with Nile blue. Later stages of gastrulæ and young plutei from such eggs, *c-f*. (After von Ubisch.)

one of the blastomeres is partly colored, the egg is removed. Most of the embryos that develop are stained on one side, indicating that the first cleavage plane coincides with the median plane. In some cases, however, the anterior or posterior side of the embryo is stained, showing that the relation is not fixed, or that it can be shifted, Figs. 32c–f.

In amphioxus the process of gastrulation, Figs. 33a–c, is very similar to that in the sea urchin, except that the inturned portion, the archenteron, is more open, producing a cup-shaped gastrula. The cavity of the blastula is obliterated when the inturning is

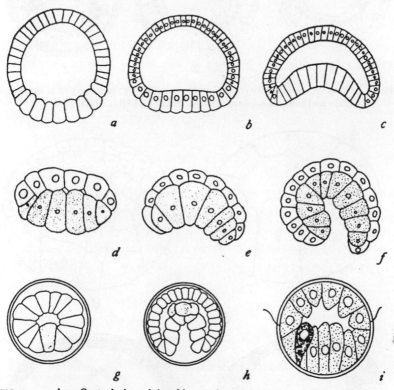

FIG. 33.   *a*, *b*, *c*, Gastrulation of Amphioxus; *d*, *e*, *f*, gastrulation of Clavellina; *g*, *h*, gastrulation of Lucifer; *i*, gastrulation of Eupomatus. (After Rhumbler.)

finished; the gastrula opening is later reduced to a small pore at the posterior end which becomes the anal opening. The dorsal wall of the archenteron pinches off to become the notochord, Figs. 34A–B. On each side of the notochord a series of pouches is formed. These pouches, pinching off from the archenteron, become the body cavi-

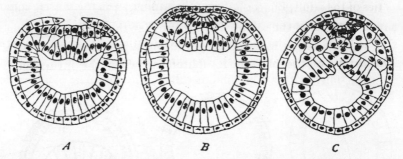

FIG. 34.   Cross sections of embryos of Amphioxus, showing the formation of the neural plate and tube, notochord, and gut pouches. (After Cerfontaine.)

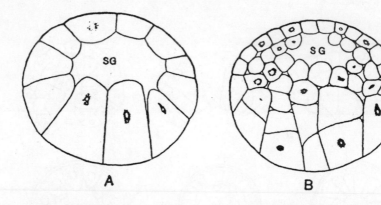

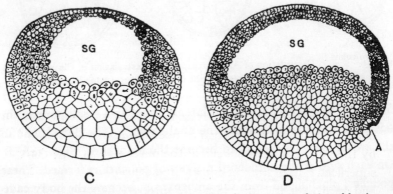

FIG. 35.   Sections of late cleavage stages, and beginning of gastrulation of frog's egg.

ties of the adult, Fig. 34C. From their inner walls the muscle somites are formed that constitute the bulk of the adult animal.

In frogs and salamanders the blastula stage, Fig. 35A, has a thin wall over the polar hemisphere with a large segmentation cavity inside, filled with an albuminous fluid. The lower hemisphere con-

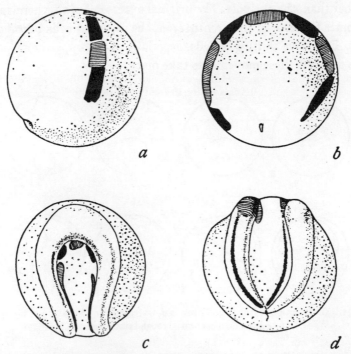

FIG. 36. *a, b,* Young gastrula stages of Triton that have been marked by stains of two colors in a half ring extending nearly halfway round the egg; *c, d,* the subsequent position of these marked areas. (After Goerttler.)

sists of large yolk-laden cells. Invagination begins at one side near the equator where the gray crescent lies, Fig. 35D. The inturning is in the form of a slit rather than an open mouth, and is brought about by the pulling inward of cells of the outer wall in that region. The superficial cells in front of and at the sides of the slit move toward it and turn over its edge into the interior. Experiments made by marking the outlying cells with dyes, Fig. 36, show that extensive regions of the outer surface move into the interior by rolling in over the edges of the opening to form the upper and lateral walls of the archenteron. The rim of the blastopore in the form of a crescent moves over the surface of the lower hemisphere, Fig. 37, as far as, or farther than, the antipole. The original outer wall of this hemisphere becomes the floor of the archenteron; the inturned cells become the roof. Extensive shifting of cells, especially of the upper hemisphere. goes on while these movements take place.

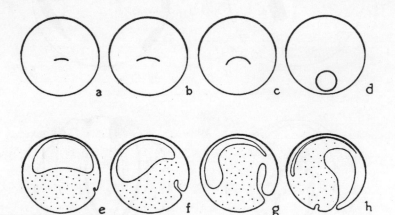

FIG. 37. Gastrulation in frog or in Triton. *a-d*, Overgrowth of dorsal lip of blastopore; *e-h*, median section through same stages.

From the cells lying along the mid-dorsal wall of the archenteron the notochord develops, and from the cells on each side of this the mesoderm or middle layer develops that gives rise to the muscle somites, body cavity etc. Both are formed in essentially the same way as in amphioxus, Fig. 34, except that pouching does not appear. The mouth is formed at the point where the anterior end of the archenteron meets the outer layer, which turns in to form the lining of the mouth. The posterior opening of the archenteron, or gastrula opening, becomes the anus.

In the vertebrates the central nervous system is a conspicuous feature of the early development. It appears in the embryo just after the gastrulation process is completed. The layer of superficial cells, lying in front of the blastopore, begins to draw together in the shape of an elongated plate—the neural plate. The edges and the anterior end of the plate begin to lift up, Fig. 36, and roll over to

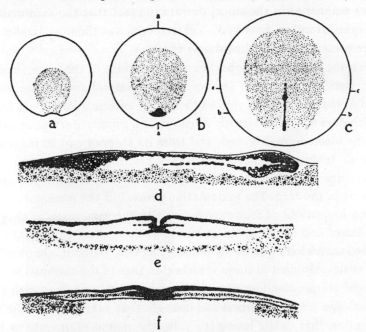

FIG. 38. *a, b, c,* Stages in development of blastoderm of chick; *d,* longitudinal section through *b; e,* cross section through primitive streak of *c; f,* cross section through notochord of *c.* (After Patten.)

form a hollow tube. The walls of this tube form the brain and nerve cord from which nerve fibers grow out to all parts of the body. The nose, and parts of the eyes and ears, are formed by inturning of the superficial cell layer of the head region.

The embryo of the bird develops from a disc of cells at the pole of the egg, which in time spreads over the large yolk mass. The first

superficial indication of the gastrula appears at the posterior surface
of the disc, Figs. 38a–b. Later the blastopore is represented by a
streak of cells called the primitive streak, Figs. 38c, f, but before the
primitive streak has appeared a plate of cells has pushed under the
disc and this gives rise to the endoderm, Figs. 38b, d.

In mammals the gastrulation takes place apparently in much the
same manner as in the chick, despite the fact that the mammalian
egg is small and without yolk. Yet it develops as though a large yolk
were present as in the lizard and the bird. The disc of cells, out of
which the embryo develops, lies on the surface of the large vesicle,
Fig. 29a. The peripheral cells of the disc extend around the inner
wall of the vesicle, Fig. 29b, as though to enclose a large yolk mass
which does not, in fact, exist. Meanwhile a primitive streak appears
on the blastoderm, Fig. 29d, and from its anterior end an ingrowth
of cells to form the notochord takes place. The archenteron and
mesoderm arise from the under layer of the disc, in much the same
way as in the bird. The gastrulation process in the mammal shows
many indications of following essentially the same path as that of
the lizard and the bird, but at the same time seems to have abbre-
viated somewhat these gastrulation processes. Without a knowledge
of the development of these vertebrates, that of the mammal would
appear unique standing alone in the vertebrate series, but with this
knowledge we can interpret the changes that take place on the assumption that, while losing its yolk, the mammalian embryo has
retained the method of gastrulation followed by its ancestors. The
overgrowth of a fictitious yolk mass to form a "yolk sac," Fig. 31b,
is essentially the same as in birds, Fig. 31a. This sac is finally drawn
into the archenteron as in birds. It furnishes a striking example of
an ancestral and now useless procedure—the yolk sac is a rudimentary organ.

There have been numerous attempts to find a mechanical explanation of gastrulation, and, in a broader sense, of the phenomenon of
invagination that plays an important rôle in embryonic develop-

ment. These explanations have rested almost exclusively on analogies with mechanical models and very little on experimental results with living eggs. One fact of observation, however, has furnished the basis for most of the hypotheses. When a layer of columnar cells turns in to form a cup, the inner ends of the cells become broader and the outer ends narrower, Fig. 33c. In other words: the individual cells become truncated wedges. Now, if over a part of the elastic outer wall of a blastula the cells should change their form so as to become wedge-shaped, the wall would automatically turn in. On the other hand, if the wall were pushed in from the outside, or sucked in from the inside, the cells would assume the same shape, since the inner surface becomes larger than the outer surface. The observed changes in the shape of cells would be the same in either case. There are, however, certain considerations that seem to eliminate the latter hypothesis.

For example: there are no known external forces that could be appealed to that would push the thicker part of the wall into the interior. If any such agent were acting, the smaller cells of the roof of the gastrula would be the ones to sink in. If the fluid in the blastula cavity were withdrawn, or squeezed out, or absorbed by the walls again, it would be the roof that would sink in, rather than the thicker floor.

An alternative hypothesis suggests that the larger cells of the wall of the blastula become broader, either because they absorb water from the blastula cavity, or because the surface tension on the inner surface is decreased. The first view fails to explain why the swelling is confined to the inner side of the cells, i.e., why the whole cell does not swell up and become spherical. Of course the inner and the outer surfaces are probably different. It might seem that the inner part would absorb water faster than the outer part. But it is not clear why such a change should take place at this particular period rather than earlier.

Several kinds of models have been constructed that turn in when the arrangements are such that the inner surface of one part of the wall absorbs fluid from the interior faster than the outer wall, Fig. 39. Turning in takes place under these conditions, but this does not prove that the same kind of differences are present between the inner and the outer surfaces of the blastula wall.

Perhaps the most plausible hypothesis is that the turning in is the result of a difference in the surface tension of the inner surface of

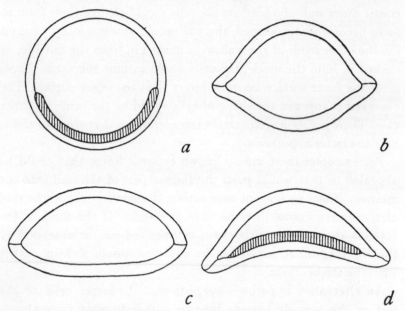

FIG. 39.   Model of gelatin cups to imitate gastrulation. (After Spek.)

the blastula wall, due to the accumulation of carbon dioxide or some other substance in the interior. As a result of the lowering of the surface tension, a movement of the materials of the cells toward that side will take place; their inner ends will become broader, and the inturning would mechanically follow.

No one of these hypotheses suffices at present, because of the lack of information concerning the nature of the cells, but they suffice at least to suggest that the process of gastrulation is not beyond the range of physical explanation. As in so many other situations in development, the analysis leads back to the behavior of individual cells; but the observed fact, that all the cells of a given area act together toward a common end, encourages one to look for an explanation that treats the initiation of the change as a reaction in response to agencies outside the cells, rather than to some change taking place in the cells individually. The latter alternative is not, however, excluded if it could be shown that at this particular moment of development the cells in a given region undergo individually an internal change that makes them responsive to the physical conditions of the system at this time. These changes might be thought of as the culmination of changes in the protoplasm that began with fertilization; or they might be thought of as new changes that the genes initiats at this time.

CHAPTER VI

HALF AND WHOLE EMBRYOS

In the classical experiment of Wilhelm Roux (1883), around which for many years a great deal of discussion took place, one of the first two blastomeres of the frog's egg was injured with a hot needle. The other blastomere developed into half an embryo. Roux drew the very natural inference that the parts of the embryo are laid down from the first division onward by differential divisions. He went a step further and referred the source of this differentiation to a qualitative sorting out of the chromatin material. Later Weismann based his entire conception of the development process on the same assumption; although, even at that time, there was abundant evidence that in the early division of the egg, as in all later divisions, each chromosome splits lengthwise into exactly equal parts. This last conclusion is amply borne out by more recent evidence from genetics that carries, in fact, the evidence beyond the visible division of the chromosomes into the invisible division of the genes themselves.

A few years later Hans Driesch (1891) carried out experiments with the eggs of the sea urchin and found that, if the first two cells are shaken apart, each develops into a whole embryo of half size. The result stood in striking contrast to that of Roux, and seemed to lead to the opposite conclusion, namely, that the differentiation does not take place at the first division of the egg. Each of the first two cells, Driesch said, is "totipotent." But the two experiments were not identical. In the frog's egg the injured blastomere, still alive, remained in contact with its fellow, while in the sea urchin's egg the two blastomeres were completely separated. The difference should not essentially affect the conclusion that Roux drew, for, if differentiation begins at the first division, this should hold for both kinds of eggs.

Since these early experiments and their interpretations, there have been many other similar experiments carried out not only with the eggs of the frog and of the sea urchin, but with numerous other eggs. Eefore discussing the relative merits of the controversies that were based on these observations of Roux and of Driesch the later experiments must be considered.

EXPERIMENTS WITH SEA URCHIN EGGS

When the first two blastomeres of the sea urchin egg are separated, each rounds up. At the next division each divides exactly as

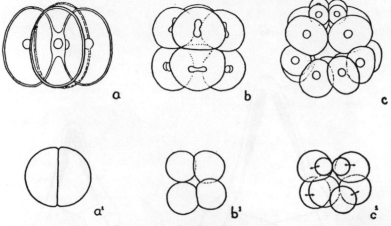

FIG. 40. *a, b, c,* Four, eight and sixteen-cell stages of egg of sea urchin; *a', b', c',* first three cleavages of an isolated half blastomere. (After Driesch.)

it would have done had it remained in contact with its fellow, i.e., through the pole, Fig. 40a'. The next division is equatorial and the next one gives rise to the micromeres at the antipole, Figs. 40b'–c'. As the divisions continue a fluid space appears in the interior, the open side of the half blastula having closed during the early divisions. Gastrulation then takes place and later a typical pluteus is formed, Figs. 41b–b .

The method of closure of the blastula and the method of invagination of the archenteron present some points of theoretical interest.

These processes can be studied by coloring the antipolar region of a half of the sixteen-cell stage with a dye. The results show that the

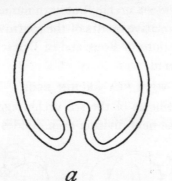
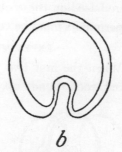

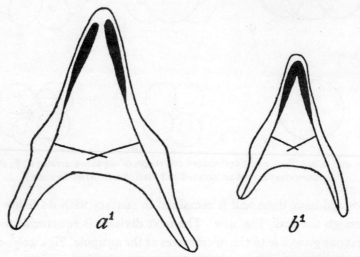

FIG. 41. Whole and half gastrulas and plutei of sea urchin; the latter from half blastomeres.

open side of the half blastula closes, the polar and the antipolar cells come together (as well as the cells from side to side). The inturned archenteron begins at the middle of the endoderm field. These relations are shown in Fig. 42, in which a whole blastula is drawn in b, its normal method of gastrulating in a, and the gastrula of the half gastrula in c. The inturning is symmetrical. Only endodermal material is turned in. This would seem to mean that the original center of the endoderm material (that after the separation lies at the side of the half blastula) is not the center of gastrulation of the half blastula, but that this center is reëstablished in the middle of the endo-

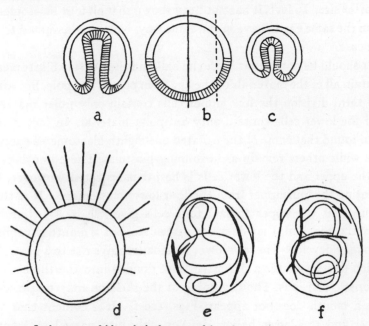

FIG. 42. In b, a normal blastula is drawn and in a its method of invagination. In c, for comparison, the invagination of a half blastula is represented. (After Hörstadius.)

dermal material. It is possible to picture this shift of center as follows. The cells at the pole have come into contact with the cells at the antipole at what was originally the center of the endoderm. If now, owing to this contact, the endoderm cells lose some of their diffusible materials and conversely the ectoderm cells give up to the endoderm some of their materials, the most concentrated material of the endoderm will come to lie nearer the middle of the endoderm mass. A new center will be formed there which becomes the center of gastrulation.

If at the four-cell stage of the sea urchin egg the four blastomeres are isolated, each segments as a part of a whole. There is formed only a single micromere at the second division. A blastula of quarter size develops that gastrulates and produces an embryo, a pluteus of quarter size. In fact, it has not been shown that all four blastomeres from the same egg behave in the same way, but this is assumed to be the case.

It should be noted that both the half and the quarter blastomeres contain all of the materials of the egg from pole to antipole, but after the third division the four upper cells contain only polar material and the lower cells contain only antipolar material. In fact it has been found that some of the isolated one-eighth blastomeres gastrulate while others remain as swimming blastulas. The potentialities of the upper and the lower cells is best demonstrated, however, by isolating the four upper from the four lower blastomeres. It is then found that the upper (polar) four cells give rise to a swimming blastula, Fig. 42d, that does not gastrulate, but a mouth invagination may develop. The four lower blastomeres give rise to a gastrula, which develops later a digestive tract divided into the three characteristic divisions. The beginning of the skeleton may be present, but a mouth does not appear, Figs. 42e–f. It is evident that the upper and the lower halves develop only into partial structures that are similar to the parts that the same regions of the whole blastula give rise to.

EXPERIMENTS WITH ASCIDIAN EGGS

In the ascidian the situation is different in some respects from that of the sea urchin, for while in both the cleavage is strictly partial, the gastrulation begins at an earlier stage in the ascidian and is, at first, more nearly a half process that only later rectifies itself to some degree.

By shaking the segmented eggs it is possible to kill one of the blastomeres without injuring the other. In this case the injured blastomere is actually killed, and not simply prevented from developing, as in the frog egg, although this may also happen in some eggs. More recently the blastomeres have been completely separated and it has been shown that they behave in the same way as when one of them is killed. Cell for cell the isolated half blastomere

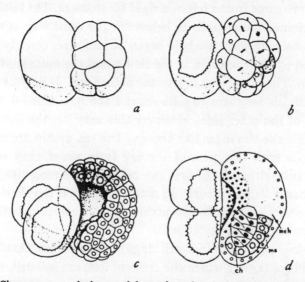

FIG. 43. Cleavage, gastrulation, and formation of embryo from half blastomere of Styela. The injured blastomere is not dead, but its development is delayed. It is still in contact with the more developed half. (After Conklin.)

goes through the same divisions as it would have done had it remained in contact with the other living blastomere, Fig. 43. It contains therefore just half as many cells as does the whole embryo at the time when it begins to gastrulate, but this takes place, so to speak, on one side instead of all around the gastrula mouth, as in the whole embryo. As soon as the cells of the archenteron, notochord and muscles are turned in, and the surface cells, destined to form the neural plate, begin to draw together, a remarkable fact is observed.

The archenteron becomes, not a half sac, but a rounded whole, the notochord forms a single rod of cells that is rounded, and the neural plate forms a single structure with a median plane of symmetry. In other words, some of the organs are whole, or nearly so, and not half structures. On the other hand, the cells destined to form the muscles fail to give both right and left structures, but form only those of one or the other side of the embryo, according to whether the embryo came from a left or a right blastomere. The failure of the predetermined muscle cells to behave in the same way as the other cells, namely, to form a whole organ of half size, can plausibly be explained by their position at the time when the notochord cells are turned in. They lie to one side of the notochord. It may be mechanically difficult for them to pass around the notochord to form the muscles of the other side. However this may be, the fact remains that, while the cleavages like those of the sea urchin are strictly in accordance with that part of the egg from which they come, the embryo from either of the first two cells is much more like that of a whole than a half structure. In many respects it develops in much the same way as does the sea urchin. This involves in both cases a large amount of self-regulation.

If the two cells of the four-cell stage of the ascidian, that lie on the dorsal side of the egg where the crescent lies, are isolated, they continue to segment as though part of a whole, Figs. 44a, b. The embryo gastrulates, but after closure of the blastopore, its progress comes to an end. The neural plate of cells is present, full size, but does not roll in to form a tube. The notochord cells are present, but do not elongate to make a rod. An archenteron is also present. The failure to elongate may account for the static condition of these organs. The embryo lacks muscle cells, caudal endoderm, and mesenchyme cells.

The other two blastomeres opposite the crescent, when isolated, cleave as part of a whole, Figs. 44c, d. Muscle cells are later present. Caudal endoderm, and mesenchyme cells are also present. There is no trace of notochord or of neural plate, and no tail is formed. It may be said of these two kinds of embryos that the one fails to develop into a whole embryo of even approximate normal form because in the absence of the other half it fails to elongate, and the other fails because it does not receive a contribution (notochord and neural plate) from the other half as does this region of the normal embryo.

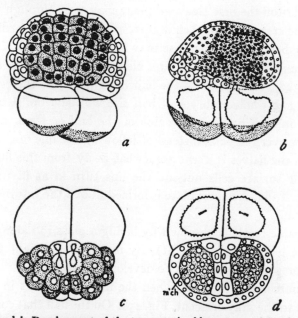

FIG. 44. *a* and *b*, Development of the two anterior blastomeres of Styela; *c* and *d*, development of the two posterior blastomeres. (After Conklin.)

The cleavage of the single quarter blastomere depends on what part of the quadrant it comes from, being in either case strictly partial. Its further development is similar to that of the two combined blastomeres just described.

HALF AND WHOLE DEVELOPMENT IN AMPHIBIA

In the light of the clear-cut results from the ascidian eggs, Roux's results on the frog's eggs can be better understood. Moreover, the experiments with the eggs of another amphibian, Triton, furnish further evidence that bears on the same problem. The first cleavage plane of the egg of Triton is usually parallel to the plane of the gray crescent, but the median plane of the embryo corresponds not with this cleavage plane, but with the second cleavage plane which passes through the middle of the crescent. In a few of the eggs of Triton the first cleavage plane may pass through the middle of the crescent, and then it corresponds with the median plane of the embryo. The latter case may be first considered. If a hair is tied around this egg in the first cleavage plane so that the egg is constricted, Fig. 45a, and if the hair is tightened as the egg further divides, two whole embryos of half size will develop. A blastopore appears in each half in the region of the half crescent, and as its lips grow over the yolk, instead of following the line of constriction between the halves it turns somewhat away from this line in each half. The surface cells outside the lips turn in as in the normal whole gastrula. A neural plate is formed above the inturned chorda endoderm.

The same result is obtained by laying a glass rod along the first cleavage plane, Fig. 46 (right). By means of a weight the rod is slowly pressed between the two developing halves.

The outcome is different when the first cleavage is in the plane parallel to the gray crescent, Fig. 45b. On the half that contains the crescent a gastrula lip develops in the crescent region. A whole embryo of half size results. The other half does not gastrulate and does not form an embryo. From these results it is clear that the presence of the region of the gray crescent, or even half of it, is essential to carry the development further.

The importance of this region is also shown by other experiments in which halves of two eggs each containing some of the gastrula lip region are grafted together. For example: if two embryos, that are each in the gastrula stage, are cut into two halves along the median axis and the right half of one half is brought into contact with the

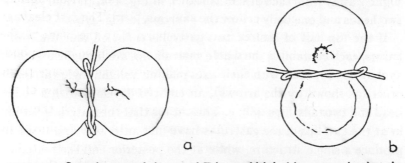

FIG. 45. *a*, Gastrula stage of the egg of Triton, which had been constricted at the two-cell stage by a hair in line with the future median plane of the embryo; *b*, a similar egg that had been constricted at the two-cell stage, but in which the first plane of cleavage was frontal, *i.e.*, at right angles to the median plane which is the more usual position of the first plane in the egg of Triton. (After Spemann.)

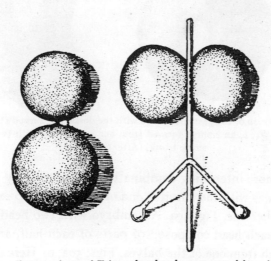

FIG. 46. Two-cell stage of egg of Triton that has been removed from its membranes (to left). The two blastomeres are separated by means of a glass rod (as shown in figure to right). (After Mangold.)

other half in corresponding position (i.e., side by side) a single embryo may result. But if, on the other hand, the halves contain the whole dorsal lip and the axes of the two united halves are turned slightly away from each other, as shown in Fig. 47a, an embryo with two heads and one body is formed, as shown in Fig. 47b.

If the top half of each of two gastrulas is cut off and the lower halves, each containing the whole gastrula lip, are brought together, as shown in Fig. 48a, with their axes making a slight angle with each other (as shown by the arrows), an embryo develops with a single head and two tails, Figs. 48b, c. This means that the materials turned in at the two lips of the gastrulas have met anteriorly and fused to produce a single structure, while at the posterior end the materials have remained apart and each formed a single structure.

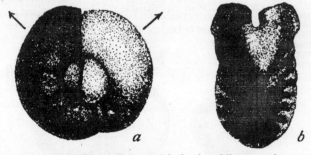

FIG. 47.   *a*, Union of two gastrula halves with the dorsal lips turned at an angle away from each other; *b*, an embryo formed from one such union with two heads and a single trunk. (After Spemann.)

A still more interesting combination is produced by uniting two half gastrulas with the anterior end of each dorsal lip exactly opposite the other one, Fig. 49a. An embryo with two heads is formed, Fig. 49b, each head composed of parts of each half, and with two bodies, each from one of the halves, Figs. 50a, b. Here the inturned materials moving forward have met at the line of union, and half of each has turned out to form one head, and the other half of each has turned out to form the other head.

The development of the frog's egg is so similar to that of Triton that it seems probable that if one blastomere were entirely removed the other would produce a whole embryo. This is the case in fact. It has not been possible to remove one blastomere without the other collapsing in most eggs, but in one of the tree frogs it was found

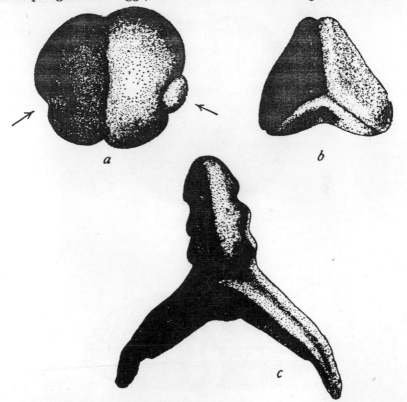

FIG. 48. *a*, Union of two gastrula halves with the dorsal lips turned at an angle toward each other, as indicated by the arrows; *b*, a neurula stage resulting from such an embryo; there is a single head formed by the union of materials from each gastrula as indicated by the color differences of the two united gastrulae; *c*, an older stage of the same embryo with a single combination head, and two independent trunk regions, one for each half. (After Spemann.)

possible slowly to suck out the contents of one of the blastomeres. The remaining one develops into a whole embryo of half size.

Indirectly the potentiality of each of the first two blastomeres of the frog's egg to make a whole embryo has been shown by inverting the egg after the two-cell stage, and holding it in that position during the subsequent stages, Figs. 51a–e. Under these circumstances it

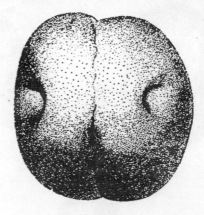

a

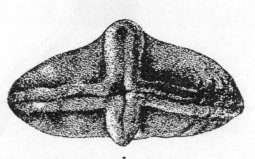

b

FIG. 49. *a*, Union of two gastrula halves of Triton with the dorsal lips directed exactly toward each other. *b*, Embryo from same; two heads are present, each formed out of materials of both halves and turned in opposite directions, and two posterior single trunk regions, one from each half. The median plane of the heads is at right angles to that of the two trunk regions. The result is a typical cruciate embryo. (After Spemann.)

is known that the materials of the interior of the two blastomeres rearrange themselves in response to gravity. The heavier yolk sinks down along the plane of division, Figs. 51b, c, pushing the gray crescent material in one blastomere away from that in the other. Each half then forms a gastrula lip, and each half produces a whole

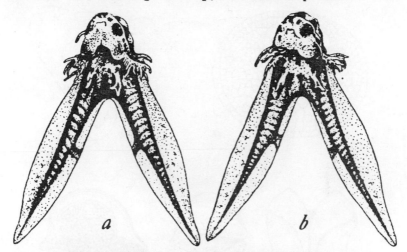

FIG. 50. Opposite sides of older cruciate embryos of Triton formed by union of two half gastrulae, as shown in Fig. 49. (After Spemann.)

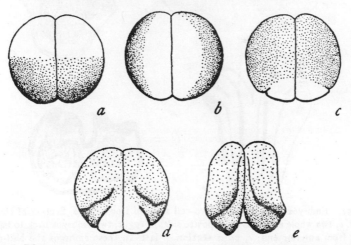

FIG. 51. Diagrams of inverted two-cell stage of frog's egg. a, Egg after inversion; b sinking down of yolk along the dividing cell wall; c, dorsal lip appearing in each half; d and e, formation of two embryos, one on each half. This method of gastrulation is not in accord with the process as described by Schleip. (After Dürken.)

embryo, except in so far as this is interfered with by the union of the halves, Fig. 51e. Several types of twin embryos, depending on the

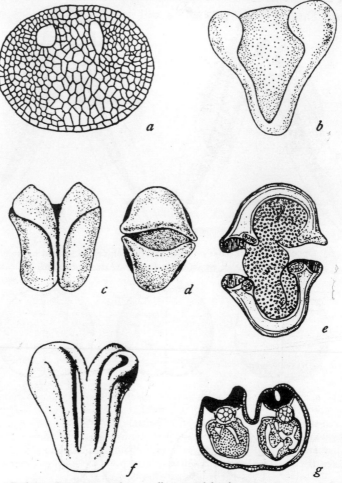

FIG. 52.   Embryos from inverted two-cell stage of frog's egg. *a*, Section of cleavage stage; *b*, two whole embryos on opposite sides of egg; *c*, two embryos back to back; *d*, same from anterior end; *e*, cross section of last. In these embryos the half-neural plates at the sides of the two embryos have not united into single neural tubes as in the typical duplicitas cruciata, and the embryos are interpreted as two whole embryos with spina bifida. *f*, Two embryos side by side; *g*, cross-section of such an embryo.
(After Wetzel.)

way in which the crescent-materials have moved apart, have been described, Fig. 52.

In the light of this evidence Roux's experiment can now be interpreted. The hot needle he used did not destroy, but only injured the

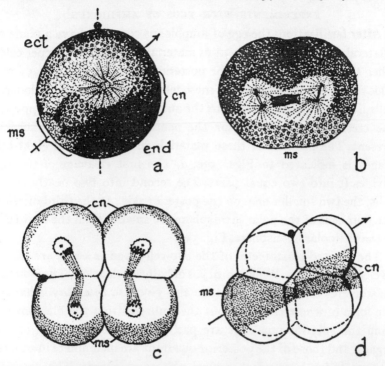

FIG. 53.   *a*, Section through median plane of egg of Amphioxus; *ms*, mesodermal crescent at posterior end; *cn*, chorda-neural crescent at anterior end; *end*, endodermal area on dorsal side; *ect*, ectodermal crescent on ventral side; *b*, frontal section just before first cleavage; *c*, end of second cleavage, posterior cells contain mesodermal crescent, *ms*. They are smaller than anterior cells with chorda-neural crescent, *cn*. *d*, Eight-cell stage, the four micromeres are shifted forward over the four macromeres—lettering as before. (After Conklin.)

materials in that part of the egg into which it was thrust. Under these circumstances the crescent region remains as a whole, half of it is turned in; hence a half embryo forms along the line of contact as it would have done had both halves developed at the same time. It might be said, perhaps, that the gray-crescent region acts as an intact organizer but only that half of it that becomes cellulated can proceed to form its half structure in the same way as it does in the normal embryo.

EXPERIMENTS WITH EGGS OF AMPHIOXUS

After fertilization, the egg of amphioxus comes to have a plane of bilateral symmetry. A crescent of material, that takes a deeper color when the egg is dyed, lies at the posterior side of the egg, Fig. 53, ms. This region of the egg becomes mesoderm. The yolk of the endoderm cells lies around and anterior to the antipole, Fig. 53, end. Opposite the crescent the material for the neural plate and notochord is present. The position of these materials in the four and eight-cell stages is indicated in Figs. 53c, d. The first cleavage of the egg divides it into two equal parts. The second into two nearly equal cells, the two smaller ones on the posterior side. At the third division the four cells of the polar hemisphere are smaller than the four cells of the antipolar hemisphere, Fig. 53d.

The isolated blastomeres of the two-cell stage as a rule are stated to cleave as halves. Whole embryos of half size develop. The isolated blastomeres of the four-cell stage also give rise to embryos having the form of whole embryos, but the embryos from the blastomeres from the anterior quadrant are partly incomplete in some of the organs, and those of the posterior quadrant are incomplete in others. The posterior larvae lack neural plate and chorda. The anterior larvae have neural plate and chorda and traces of somites, but never form complete larvae. The development of the isolated blastomeres of amphioxus is similar in many respects to that of ascidians.

CHAPTER VII

THE DEVELOPMENT OF EGG FRAGMENTS

The development of fragments of eggs, that have been cut apart before fertilization, makes it possible to carry the analysis of localization as observed in the isolation of blastomeres back into the cell itself. Here we have one of the rare opportunities in biology to study at close quarters the properties of the cell, where the regulative processes of the organism take place.

The earliest observations were made on fragments obtained by breaking the sea urchin's egg into pieces by shaking the eggs violently in a tube half filled with water. This procedure often injures the fragments, and, since the exact origin of each fragment is uncertain, the method has been abandoned. By the use of fine glass needles it is possible, under a binocular microscope, to cut eggs in two at any desired level with little injury to them. When the cut passes through the poles it is spoken of as meridional; when at right angles to this axis as equatorial.

A great many experiments have been made with the eggs of different species of sea urchins, but the results are not altogether consistent. Some of these differences may be due to differences in the methods employed. Since, in most species, the polar bodies have disappeared in the ripe eggs, and since the pronucleus shifts its position from the polar axis, the plane of the cut with reference to the egg axis is difficult or impossible to determine. The funnel in the jelly is the only means of orientation, but the presence of the jelly makes the operation difficult or uncertain. In one species, however, Paracentrotus lividus, there is a ring of pigment in the antipolar hemisphere of the egg, Fig. 14, which makes orientation of the incision possible. The following description applies, therefore, mainly to experiments with this egg.

When the egg is cut in two through the equator, i.e., parallel to and "above" the pigment ring, the polar fragment segments (after fertilization), as does this part of the whole egg, i.e., at the eight-cell stage (corresponding to half of the whole sixteen-cell stage), all the cells are equal. The opposite half, the one containing the pigment band, divides in the same way as does the whole egg, i.e., it shows the characteristic pattern of the whole sixteen-cell stage, Fig. 54a, viz. four micromeres (at the antipole), four macromeres (in the middle), and eight mesomeres (at the other end). In other words it does not divide as part of the whole but divides as does the

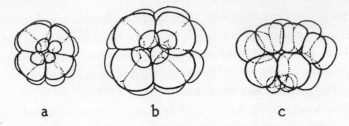

a   b   c

FIG. 54. The sixteen-cell stage of the antipolar half of the unsegmented egg. The pattern is the same as that of a whole egg at this division. (After Hörstadius.)

whole egg. The tiers of cells show the same relative proportions to each other as in the normal egg, Fig. 54b.

When the egg is cut in two through its axis, i.e., at right angles to the ring, the cleavage pattern is more variable. Most of the halves give the normal sixteen-cell configuration, Fig. 54c, especially if the egg was cut before fertilization, but others give partial types of cleavage of various patterns. The difference between the different types of cleavage of the fragments may be due, in part, to the difficulty in cutting the eggs exactly through the poles, and perhaps also to differences in the closure of the cut surfaces or to the presence of a cut surface on one side, etc. The main points, however, are fairly clear. The polar and antipolar hemispheres of the egg behave differently, and in such a way as to suggest that they contain different materials. Granting this, the proportionate type of development in the antipolar halves and to a less extent in the lateral halves can be understood to some degree since the mitotic figures that determine the planes of cleavage are regulated by the amount and distribution of the materials of the egg out of which they develop. With half the amount of the material the size of the mitotic figure should be half in the half fragment. The interpretation does not in itself suffice to explain the relative sizes of the blastomeres, for the micromeres, for example, in the antipolar halves might be expected to be too large. There must then be assumed additional factors. In fact, there is evidence that there is no sharp line between the material that goes into different blastomeres. For instance, the kind of material that goes into the micromeres is not all contained in the normal micromeres, but extends farther in the polar direction—all of it is not used for micromeres. There must be, then, some readjustment or redistribution of the relative layers before or during cleavage to give the proportionate cleavage pattern.

Fragments of the unfertilized eggs of an ascidian, Ascidiella, have been studied. At present the results are not sufficiently explicit to furnish a basis for more than rather general statements. When the egg is cut in two through or near the equator, various types of abnormal embryos are obtained. The polar half may be deficient in endoderm, and the antipolar half deficient in ectodermal parts. Pairs of embryos may be obtained, one composed of ectodern, endoderm, mesoderm, and mesenchyme; the other composed of ectoderm, endoderm, nervous system and notochord.

When the egg is cut through, or nearly through its polar axis, pairs of half embryos may be obtained, or two symmetrical embryos, or even pairs with complementary defects. It has been suggested that half embryos result when the cut passes through a predetermined median plane, and the symmetrical embryos when the cut is perpendicular to that plane. Until the finer details of the cleavage stages of the fragments and the early formation of the organs have been described, this interpretation of the results of the experiments remains very uncertain.

The development of fragments of the egg of the nemertean worm, Cerebratulus, has been studied more exactly than that of any other animal. The precise location of the incisions is possible, and the consistency of the protoplasm is such that after the operation the fragments quickly round up without obvious injury. The egg, when obtained from the female, has a rather large germinal vesicle in the polar hemisphere, Fig. 55a. At the antipole there is a prolongation

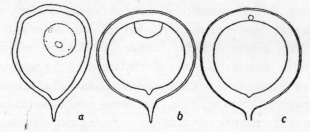

FIG. 55. Egg of Cerebratulus during the ripening stages. In *a*, the egg nucleus (germinal vesicle) is still present. In *b*, the polar spindle has formed, and in *c*, the first polar body has been given off. (After Wilson.)

of the jelly membrane, and even a short protrusion of the egg substance is present at first.

The normal egg divides equally into two, and then into four equal cells, Figs. 56a–b. The third cleavage gives nearly equal cells, but the four polar cells are slightly larger than the antipolar, Fig. 56c. This division is spiral and anticlockwise. The following divisions are alternately clockwise and anticlockwise, Fig. 56d. Later the antipolar cells are invaginated to form the gastrula. A typical free-swimming larva, called a pilidium, Fig. 57, is formed.

After the egg nucleus breaks down to form the polar spindle, the egg may be cut equatorially into two parts, and each part fertilized. The   fragment, containing the egg pronucleus, gives off two

polar bodies. The antipolar fragment does not give off any polar
bodies, since an egg nucleus is absent, but develops with only the

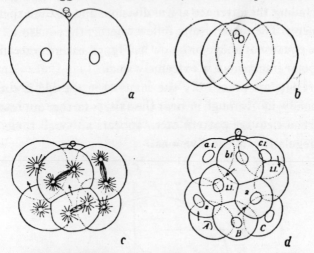

FIG. 56. Normal cleavage of Cerebratulus. (After Zeleny and Wilson.)

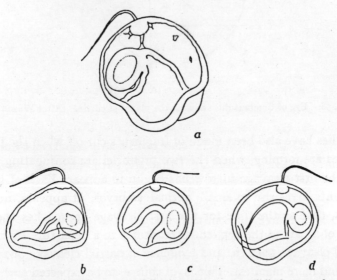

FIG. 57. *a*, Normal pilidium stage of Cerebratulus; *b, c, d*, pilidia from egg fragments
(After Wilson.)

sperm pronucleus which furnishes its "division center." The cleavage pattern of both fragments is that of the whole egg in every detail, including the alternate spiral divisions and size relations of the blastomeres. The same results follow whether the equatorial incision is at the equator or above or below it, Fig. 58, except when the polar or antipolar fragments are extremely small.

The results are practically the same when the egg is cut in two meridionally, i.e., through or near the axis, or farther out to one side. The normal cleavage pattern often appears although there may be some irregularities in the divisions.

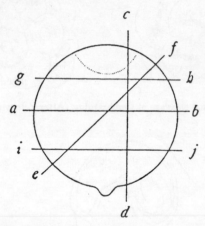

FIG. 58.   Egg of Cerebratulus showing the planes of cutting. (After Wilson.)

Studies have also been made of fragments cut off when the polar spindles are forming, when the two pronuclei are conjugating, and after the first cleavage furrow has begun to appear. Each of these fragments gives rise to small normal embryos. It appears, nevertheless, that as the time for the first cleavage approaches the removal of a part of the egg leads frequently to a departure from the typical cleavage pattern, and evidence of partial cleavage becomes more and more manifest. This is, of course, to be expected since the isolated blastomere of this egg, as in all others, gives a partial cleavage. The results here confirm the evidence from other eggs, namely, that a progressive series of changes takes place in the egg after fertilization, foreshadowing the cleavage that is about to begin. The changes are known in some cases to involve visible shifting of the materials of the egg that are connected with the subsequent events. That these movements are themselves dependent on already-present arrangements of the materials of the egg there can be little doubt, but to what extent they depend on the visible stratification of the egg's materials, or to what extent they depend on invisible physical differences in different parts of the egg, is totally unknown at present. But that the development is not dependent on a strict prelocalization of organ-forming materials of the egg as a whole is manifest from the fact that a piece will under certain conditions do what the whole egg does.

Fragments of the unfertilized eggs of several other animals have been studied, the ctenophore Beroë, the annelid Chaetopterus, the mollusc Dentalium, the chordate Amphioxus, and the vertebrate Triton, but, since these have given no results essentially different from the examples just given, an account of them may be omitted.

CHAPTER VIII

SINGLE EMBRYO FROM TWO EGGS

The study of egg fragments has shown that two embryos may develop from fragments of one egg. The converse situation—the development of a single embryo from the fusion of two whole eggs—has also been recorded. For example: if two eggs of the nematode worm, Ascaris megalocephala, fuse before the polar bodies are formed, Fig. 59, the polar bodies are later given off at the pole of each egg. If the poles lie near together the polar bodies are near together, but if the eggs have fused in such a way that the poles lie apart, the polar bodies are also given off at different points. If one sperm enters it may happen that it fuses with the two pronuclei to form a triploid embryo. Such a double egg may be supposed to give rise to a giant embryo. Giant embryos, i.e., embryos of double size, have been found. Unfortunately it is not known whether double embryos develop only when the polar field of the two united eggs lie near together or may develop even when they lie far apart. It also appears that two eggs of ascaris may be fused together after the polar bodies are given off, but whether normal embryos develop from such combinations is not known. The presence of two egg pronuclei, and the possibility of two sperms entering, introduce complications that probably would interfere with normal development.

Giant eggs of the nemertean, Lineus ruber, have been recorded. They arise from the fusion of two eggs. Fusion may occur before cleavage, or during cleavage, or in the blastula stage. Under what condition of fusion normal embryos of double size may develop is not known.

It has been found possible to produce double embryos of triton by bringing together eggs in the two cell stage, provided the orientation of the crescents of the eggs is the same or nearly so. For example: after removal of the membranes the first two blastomeres separate widely from each other. At this time one egg is laid across the other as in Fig. 60a, with both poles turned up; the four cells will unite as

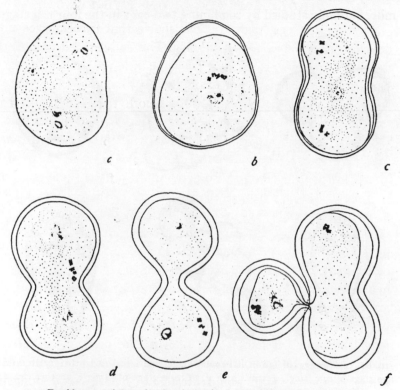

FIG. 59. Double eggs of Ascaris, showing polar bodies (diad groups of chromosomes) egg nuclei (also as diads), and sperm nuclei. (After Kautzsch.)

the blastomeres draw nearer together. Such combinations may give a single embryo as in Figs. 60c, d, or two or even three embryos. It is probable that a single embryo develops only in those cases where the crescent regions of the two eggs happen to lie near each other; if not, two or three united embryos may develop.

Numerous attempts have been made to produce single sea urchin embryos by the fusion of two eggs, either before or after cleavage. In the great majority of cases two united embryos result that are imperfect in one or more respects, Figs. 61 and 62. Rarely a single embryo has been produced, Fig. 62a. Recently the problem has been more carefully studied by combining two eggs in the four-cell stages

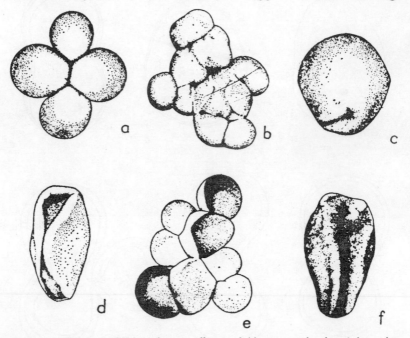

FIG. 60.   a, Two eggs of Triton, in two-cell stage, laid across each other; b, later cleavage stage of a; c, single gastrula and, d, neurula stage of same; e, another cleavage stage of two eggs united as in a; f, neurula stage of same. (After Mangold.)

in which the polar axes of the two eggs are in line. The orientation of the combination is indicated later by the location of the micromeres and by the gastrula invaginations. One of the two eggs whose membrane has been removed is first stained in Nile blue. The grafting is performed in calcium-free sea water after which the eggs tend to stick together. One of the four-cell stages is placed on top of the other and a little pressure applied. The three possible combinations are shown in Figs. 63a, b, c. When united, as in a, with the poles in the same direction, two invaginations appear, one at *anti-p.*, the other at the equator of the blastula. More rarely the gastrulation of the upper component may be suppressed, and a single digestive tract formed

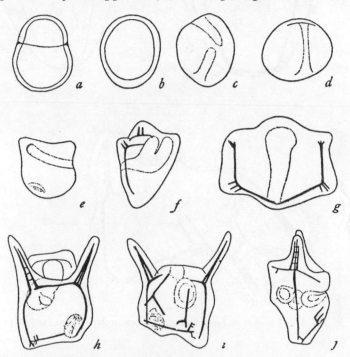

FIG. 61. *a* and *b*, Fused blastulae of sea urchin; *c, d, e*, gastrulae of same; *f-j*, plutei from same. (After Morgan.)

from the lower component. This is an approach to a single embryo. When united, as in b, two invaginations usually take place at opposite ends of the blastula. When united, as in c, there may be one or several invaginations at the equator, or invagination may be suppressed in both components. Whether the fusion of two eggs before fertilization, or immediately after fertilization and before division,

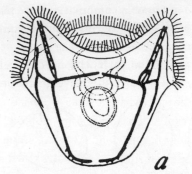

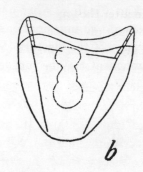

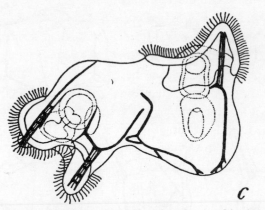

FIG. 62.   Plutei from fused blastulae of sea urchin. (After Driesch.)

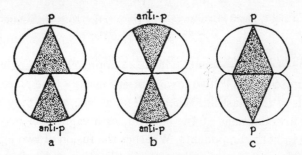

FIG. 63.   Diagram showing orientation of fused eggs of sea urchin with respect to
polar axes. (After Balinsky.)

is more likely to give, in certain combinations, more complete plutei than after the four-cell stage is completed, as above, is not definitely known at present.

Eggs with two nuclei have been found in moths. Each nucleus gives off two polar bodies. Fertilized by two spermatozoa, each nucleus divides, and together they give rise to the nuclei of the embryo. A single adult moth results. These eggs are not double-sized, and the adults are not larger than normal. At what stage the two cells have been united to produce a single egg is not known, but since these eggs are the same size as the normal egg, the union must occur before the yolk accumulates in them.

Binucleated eggs also arise on rare occasions in the fly Drosophila. These eggs have not been observed directly, but their occurrence is deduced from certain kinds of mosaic flies that have been reported. For example: when such eggs are heterozygous for one pair of chromosomes, it may happen, when the polar bodies are given off, that one nucleus loses one member of the heterozygous pair of chromosomes and the other one the other member. If the two spermatozoa that fertilize such an egg are of a kind to reveal the presence of the two reduced egg nuclei, i.e., if they contain only recessive factors, a mosaic will be formed, one part of its body showing one character and the other part the other character received from the mother. Since the flies are of normal size it is probable that such eggs arise by fusion at an early stage, or more probably by the failure of a proto-plasmic division of an oöcyte after its nucleus has divided. In cases of this sort the original polar axis of the cell would remain, and this condition would facilitate the normal development of the eggs.

CHAPTER IX

TWINS AND TWINNING

When the eggs of the sea urchin or of amphioxus in the two-cell stage are shaken, two united embryos frequently develop, owing to the incomplete separation of the two blastomeres. Each twin is half the size of a normal embryo. A study of the stages through which these twins pass shows that each develops in the same way as when the two isolated blastomeres develop, except in so far as their union interferes with such development. For a time it was supposed that all twins and double embryos originate from the accidental separation of the first two blastomeres, but later observations have made this view improbable for most cases.

A study of the development of the united twin embryos of amphioxus, Fig. 64, has shown that they arise not so much from the separation and reunion of the two blastomeres as from the shifting of the blastomeres so that their axes are turned in different directions. The polar relations are retained in each, and gastrulation takes place at the antipole of each half. If the antipolar regions remain near together a single archenteron may develop with indications of bifurcation that may or may not be subsequently rectified, but if the antipolar regions are more widely separated, as in Figs. 64a–f, two independent invaginations take place. As a result of these differences, the twins are found to be united in all possible ways, Figs. 64g–h.

Twins may be produced from the frog's egg by inverting it in the two-cell stage, Fig. 51. In this case the twinning is shown to be due to the rotation of the interior of the two blastomeres in such a way that two half crescents are separated, each behaving subsequently independently of the other, Figs. 52b–g. In principle the results are the same as in amphioxus and in the sea urchin, for it is not so much the polarity as the separation of regions of invagination

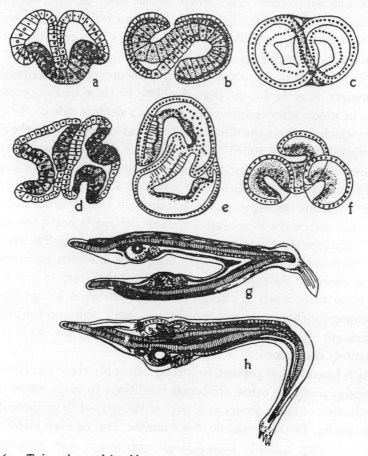

FIG. 64.   Twin embryos of Amphioxus. *a-e*, gastrulation in several embryos; *f*, triplet gastrulation; *g-h*, older twin embryos, *g* united only in tail region, ventral surfaces together; *h* united along dorsal surfaces. (After Conklin.)

that is significant. In triton, twins can be obtained by constricting the egg at the two-cell stage, or even later, in such a way that the region of gastrulation is divided into halves. In all these cases the significant feature is that each separated region becomes a new *whole of half size.*

In addition to these cases where double embryos have been obtained by partial separation of the first two blastomeres, there is another group of cases in which doubling has been experimentally induced by changing the direction of the first cleavage division, either by mechanical processes, by temperature, or by centrifuging. Moreover, it is significant that doubling in these cases occurs in eggs in which, after fertilization, there is a definite side of the egg from which originate the mesoblast cells that appear to play the rôle of organizer. In the annelids, Chaetopterus and Nereis, and in the mollusc Cumingia, the first division is into unequal parts. The larger cell contains the future mesoblast material. If the egg is put under pressure before the cleavage, the first division is equal in many eggs. In consequence the mesoblast region of the egg is now present in both blastomeres. These eggs produce double embryos. The results are most important because they bring into line with the normal development of the eggs the particular event that leads to doubling. In other cases where a change in the environment brings about doubling—such as temperature changes, both high and low; ultra violet, etc.—the effect of the agents in question has not as yet been definitely determined.

It is hazardous at present to attempt to apply these results with artificial media or other abnormal conditions to cases where the production of twins occurs as a part of the normal development of the species. In the armadillo, for example, four or even eight embryos develop regularly from a single egg; and in man, also, twinning is of rather frequent occurrence. A tendency to produce human twins seems also to be inherited. Before considering the latter some environmental effects may be described.

In the bony fishes, especially in trout, double embryos are occasionally found. When first seen they lie at two separate oints at the edge of the blastoderm, Fig. 65. If they are opposite each other the

two resulting embryos become later united by their ventral surfaces, as in Fig. 66. If they lie near together they may coalesce as
the blastoderm grows over the yolk and form a two-headed twin with
a single body, as in Fig. 67. Sometimes one embryo may be defec-

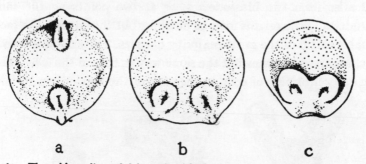

a   b   c

FIG. 65. Three blastodiscs of fish, each with the beginnings of two embryos. (After
Rauber.)

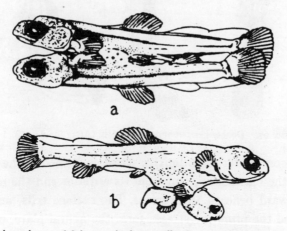

a

b

FIG. 66. Twin embryos of fish. *a*, united ventrally; *b*, one embryo much smaller than
other. (After Stockard.)

tive or smaller than the other one, Figs. 66b and 67c. These types are
clearly due to the gastrula ingrowth pushing under the blastodisc at
two separate points, their subsequent union depending on the overgrowth of the rim of the blastoderm. What intrinsic or extrinsic

factors are responsible for the initiation of the two-fold invaginations is not known.

In birds, especially in the fowl, many double embryos have been described, and there has been a good deal of discussion as to whether they arise from two blastoderms, or at two points on the same blastoderm. Both origins may be possible, but the latter is the more probable origin of the great majority of cases. The early embryo of the chick arises in somewhat the same way as that of the fish. There is an early inturning of cells at the edge of the blastoderm to form

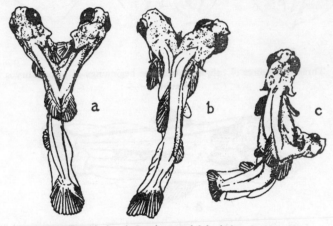

FIG. 67.   Double-headed embryos of fish. (After Stockard.)

the endoderm. Later the primitive streak is formed in the posterior region of the blastoderm, and from its anterior end the notochord grows forward beneath the surface. The surface cells, anterior to and around the primitive streak, form the neural plate, etc. Some of the twin embryos of birds have two heads and a single trunk, Fig. 68a; others have one head and two trunks, Fig. 68b; less often they are entirely separate from each other, Fig. 68c. Not infrequently the two heads may be in contact, the two trunks extending outward in opposite directions. In the last case it appears that the twinning, if it may be called such, is due to two independent begin-

nings, but the cause of the origin of two developing centers is not
known. It is to be observed that when the head or tail ends are
separate, they each have a right and left side, but when they unite
to form a single structure there is only one right and one left side. In
other words, the axial organs, nerve chord and notochord form

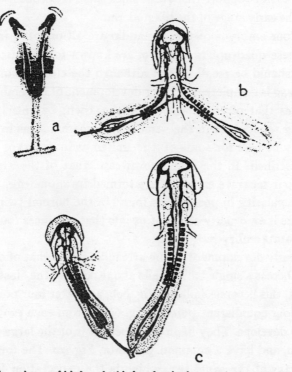

FIG. 68.   Twin embryos of bird. *a*, double head united at trunk; *b*, single head, double
body; *c*, two whole embryos in one egg. (After Tannreuter.)

single median organs, and there is only one right and one left side
consisting of the mesoblastic blocks and other right-left organs.
This is much the same phenomenon that appears in the Janus embryos formed by uniting two gastrula halves of the eggs of triton,
described in Chapter VI.

The small egg of mammals, Fig. 28, divides completely into two, four, eight cells, etc. In this respect the cleavage is more like that of amphioxus or sea urchin, and it might be supposed that the separation or shifting of the first two or first four cells explains the origin of two embryos from the same egg. But the evidence, as far as it goes, does not support this view, since there is no observable doubling in the early stage of the blastoderm.

The four embryos of the armadillo are all oriented in the same way. These quadruplets, Fig. 69, are known to arise from a single egg. It should be recalled that although the cleavage of the mammalian egg is complete, the later development of the embryo closely resembles that of the chick. It might, then, be anticipated that twinning would arise in the same way in mammals as in birds, and this may be true for some of the united twin embryos that have been described. In the best known case, that of the armadillo, in which four separate embryos are formed from one egg, there is no such irregularity in position as found in the normal twins of birds, but there is an orderly series of events that produces four complete and separate embryos.

The early development of the armadillo is like that of other mammals. There is a single ball of cells at one point of the blastocyst wall, Fig. 69a, this becomes hollow, Fig. 69b, and later four pouches grow out at four equidistant points, Fig. 69d. From each pouch a single embryo develops. They lie in four meridians of the large blastocyst, Fig. 69c, and have a common placenta, Fig. 70. The four embryos are always of the same sex, which is the expectation for one egg fertilized by one sperm. The sex-determining mechanism of the mammal, XX-XY, explains why all four embryos are males, or else females. It has, in fact, been shown that only one egg at a time is set free from the ovaries.

The term twin applied to man means two babies born at the same time. Most twins come from two separate eggs, each fertilized by different sperm; hence they may be two boys, or two girls, or a girl

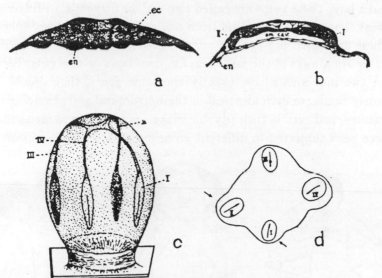

FIG. 69. *a*, Section of early blastoderm of Armadillo; *b*, later stage with amniotic cavity present; at I and II lateral pockets are pushing out in each of which an embryo will develop; *c*, a later stage with four embryos; *d*, a surface view showing position of four embryos. (After Patterson.)

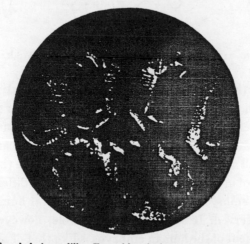

FIG. 70. Nine-banded Armadillo. Four identical twins with a common placenta. (After Newman and Patterson.)

and a boy. These twins are called two-egg, or dizygotic, or fraternal twins. The other twins come from one egg fertilized by one sperm. These are called one-egg, or monozygotic, or identical twins. The latter are always of the same sex, i.e., two boys or two girls. Since the two individuals have exactly the same genes, they should be closely similar or even identical in their structural and physiological features, and even in their psychic make-up, except in so far as they have been subjected to different environments. As a precaution it

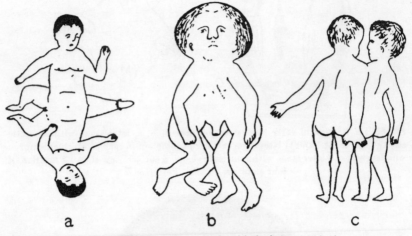

a   b   c

FIG. 71.   Human twins united at birth. (After Wilder.)

should be further stated that any irregularity of location of the two embryos with respect to the placenta or to each other might lead to one twin developing faster or better than the other, and this embryonic environmental difference might give one twin an advantage, however small at the start. With an initial difference the subsequent history even of identical twins might be somewhat different.

The early embryonic development of identical human twins has not been observed. It seems not improbable, however, that since the development of the human embryos is like that of all other mammals, human twins arise in somewhat the same way as do the quadruplets of the armadillo. In both there is some evidence of mirror symmetry which indicates that they have arisen from a common vesicle, and reflect in their right-leftness a relation to each other. In this connection it should be noted that identical twins are not always separate at birth. There are many reported cases where the twins are united in all possible ways, Fig. 71. Most of these die at birth, but some of them remain alive. The Siamese twins are the classical example of twins of this kind.

Identical twins show similarities of structures down to their smallest details, even in such characters as their finger prints. This gives a measure of the strength and extent of the hereditary units. It would be difficult to invent an experiment more satisfactory than this to establish the influence of the genes on the somatic characters. It is true that both embryos develop under identical environmental conditions, but so do fraternal twins that develop from separate eggs fertilized by different sperm; yet the latter are no more alike than are other children of the same parentage born at different times.

For the study of human psychology the occurrence of identical twins offers very rich opportunity to discover whether their psychic equipment is also identical at the start, or whether the laws of inheritance are not applicable to this field. There is an ever-increasing number of recorded cases where babies that are identical twins have been reared apart. Their structural resemblance remains much the same as when reared together, but with respect to their psychological resemblances and differences, there is at the moment no safe statement that can be made, unless we may venture to say that on the whole environment plays a less significant rôle than heredity—a conclusion that Galton reached years ago in his study of twins. There can be little doubt, however, that with improvement in the tests of psychological behavior that will enable us to discriminate better than at present between the innate capacity to learn and the external effects of social relations and education, the prospects of finding out how much is due to nature, and how much to nurture, are very promising.

CHAPTER X

MULTIPLE CHROMOSOME TYPES

The early work in experimental embryology led to the conclusion that at least one set of chromosomes, i.e., one of each kind, is necessary for normal development; but, as has been pointed out, this conclusion left open the question to what extent additional chromosomes would affect the result. That a single set of chromosomes alone may suffice was shown by the development of normal embryos from non-nucleated fragments of eggs fertilized by a single spermatozoön. This evidence demonstrated, experimentally, a fact already known in respect to certain insects, namely, that the males of rotifers, of some species of bugs, and of bees start with the half number (haploid set) of chromosomes. In the latter cases the chromosomes are derived solely from the nucleus of the egg that develops parthenogenetically, i.e., without being fertilized. More recent work in genetics has added a great deal of information as to the effects of adding one or more chromosomes or parts of chromosomes to the normal number, and has also shown what happens when the chromosomes are three times (triploid), four .times (tetraploid), or several times (polyploid) the basal number (haploid) of chromosomes of the species.

Three main problems are involved. First, the relation of a change in chromosome numbers to the amount of protoplasm that develops under such conditions. Second, the problem of balance between the genes. Third, the problems that arise when the maturation process takes place, i.e., when like chromosomes conjugate with each other.

Occasionally two sperm nuclei, from two spermatozoa that have simultaneously entered an egg, fuse with the egg nucleus, giving three sets of chromosomes. But since the sperm nuclei have each brought a centriole into the egg, which divides giving four poles to the spindles, Fig. 2, a complicated mitotic figure develops, and the distribution of the chromosomes is erratic. These cases furnish poor material for the analysis of the present problem. A much better chance for interpretation is found in plants when a pollen grain, which rarely happens to carry two sets of chromosomes, fertilizes an egg with one set, or vice versa. In plants the centrosome difficulty does not arise. A triploid individual results. Less often, cases are found where the egg and the pollen grain both carry a double set of chromosomes, giving an individual that is tetraploid.

Now an animal egg fertilized by a diploid spermatozoön is expected to divide the usual number of times, and the effect of the chromosomes on the volume of the protoplasm, if any, is not expected to appear during these stages of development. But in time this influence prevails, and the cells become larger than those of the normal diploid. Similarly for a tetraploid, where it has been shown in several cases that the individual cells of the adult are twice the volume of the normal individual. Evidently the additional chromosomes affect, in time, the amount of protoplasm produced.

The characters of the tetraploid individual are essentially the same as are those of the diploid, or ordinary individual from which it was derived, but in some other respects, such as the size of the individual or the proportion of the parts, the tetraploid may be different. Some differences are to be expected because of the relation of volume and surface; doubling the volume increases the surface only 1.59 times. Whether the total number of cells is the same in diploid and tetraploid individuals has not been accurately determined, but it seems probable that the total number is the same in both, or nearly so. But this may be different in different organs of the body.

In other cases where the double number of chromosomes in an early stage of the egg or in a somatic cell arose through the fusion of the halves of a cell that is about to divide (whether a young egg cell or a somatic cell is immaterial), the protoplasm may be considered as having doubled. In such cases the problem of the relation of chromosomes to protoplasm, spoken of above, when a diploid egg that has reached its full size has its chromosomes suddenly doubled or trebled, does not arise.

Information in regard to the converse situation, namely, the relative size of the cells in the haploid individual, has been reported in a

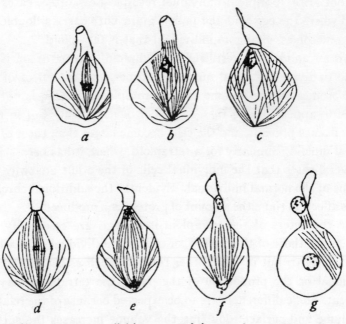

FIG. 72.   First spermatocyte divisions, *a-c*, and the second spermatocyte division, *d-g*, in the bee. (After Meves.)

few cases. The size and number of the cells of the male bee, that starts as a haploid, have been examined. In the adult the sizes of the cells, organ for organ, appear to be the same as those of the diploid female, but there is some evidence—inadequate, to be sure— showing that in the male the chromosome number may become doubled in the early stages of development. If doubling of some sort

takes place in the body cells, it does not take place in the germ cells
of the male as shown by the two maturation divisions, where the
haploid number of chromosomes is demonstrably present, Fig. 72.

In mosses and ferns and to a lesser extent in seed plants, there is a

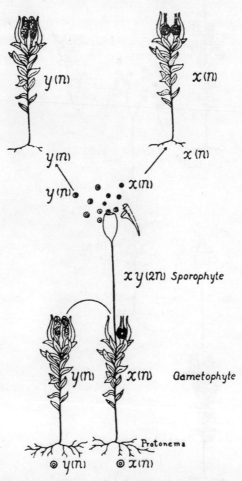

FIG. 73.   Life cycle of moss. The mycelial thread and the moss plant constitute the *1n*,
or gametophyte generation; and the stalk and capsule (with its contained spores, arising after fertilization out of the moss plant), constitute the *2n* or sporophyte generation.
(After Marchal.)

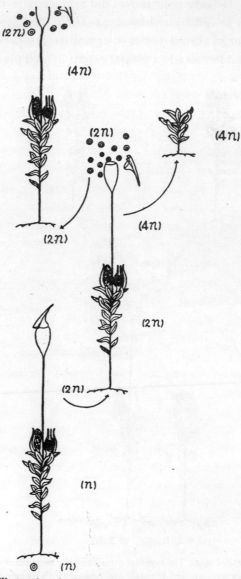

FIG. 74.   Diagram illustrating the formation of *2n* individuals from the regeneration of the sporophyte in a hermaphroditic species. (After Marchal.)

haploid generation or gametophyte, and a diploid one or sporophyte. They alternate in the life cycle, Fig. 73. The two stages are so different that a comparison of the relative cell sizes is not instructive. However, in mosses it has been possible to double artificially the number of chromosomes in the haploid generation. For example: if a piece of the stalk of the diploid generation (the sporophyte) is cut off and placed on wet sand, the cells grow out and produce the gametophyte generation, Fig. 74. Its cells contain the diploid number of chromosomes, since the cell from which it came was diploid. The same process may be repeated with the tetraploid sporophyte that comes from the last, and from which a tetraploid gametophyte is produced. The cells are approximately twice as large as the diploid, and four times as large as the normal. The relative sizes of the cells in the diploid and tetraploid gametophytes (protenema), as well as the size of the cells in the organs of the sporophyte have been studied. In both, the cells are larger, approximately in proportion as the number of chromosomes is multiplied.

The last examples show clearly that, in the presence of a double or quadruple set of chromosomes, the amount of the protoplasm is correspondingly increased. The experiments show also that the transition from sporophyte to gametophyte is not due to the presence of one set or two sets of chromosomes.

The second question concerns the balance of the genes. This expression is intended to imply that the total activity of the genes is exactly such that it gives the characteristics of the individual. This may seem a platitude, but a real meaning becomes apparent when considered in relation to a change in the number of genes. Two kinds of changes are characteristic: first, the total diploid number may be multiplied or halved. The balance is then the same as before, and a normal product is to be expected. But if one, two or more chromosomes are added to a complete set, the balance is changed. There are then relatively more genes of certain kinds than before. Many examples of this unbalance are known.

For instance: owing to some accident in the division of a germ cell (or of any other cell) a particular chromosome may fail to send its halves to opposite poles, and as a result one daughter cell comes to contain an extra chromosome; the other to lack that chromosome. Should this happen, as it does at times, in a germ cell, and the chromosome deficiency or duplication be carried into the fertilized egg, an individual is produced that is "unbalanced." It may survive, and if so it generally shows recognizable differences from the original type. On the other hand, such an individual, whether plus or minus for one chromosome, may die because the unbalanced relation of the

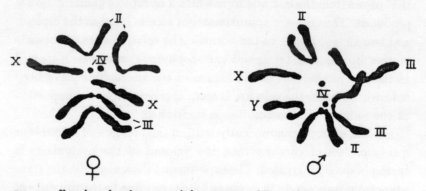

FIG. 75.   Female and male groups of chromosomes of Drosophila melanogaster. (After Dobzhansky.)

genes leads to irregularities in its development. For example: Drosophila melanogaster has four pairs of chromosomes, Fig. 75, three of them much larger than the fourth pair. An absence of one of the smallest chromosomes, or the addition of a third one to an individual affects the individual so slightly that it resembles the normal fly in all but a few slight differences. But the absence of one of the larger chromosomes, or the addition of a third one affects the development so adversely that the embryo dies.

An apparent exception appears in the case of the sex chromosomes. In some forms the female has two such X-chromosomes and the male only one, yet males and females are very much alike in nearly all their characters. In Drosophila, the female has two X's and the male one X, but in addition another sex chromosome called the Y. It has been shown that the Y lacks nearly all of the genes present in the X. In such cases as these it may be said that some sort of compensating balance must have been established, so that one X and two X's give similar results—except, of course, in so far as two X's give a female and one a male, which differences are due to a difference in balance between the X's and the rest of the chromosomes.

Further examples illustrating the balance of the genes in relation to sex are as follows: a tetraploid female of Drosophila has the formula 4A + 4X, and the normal diploid female 2A + 2X. The letter A stands for all the other chromosomes, except the X. The balance is the same in both. A tetraploid female mated to a normal male (2A + X) gives triploid females (3A + 3X). Here also the balance is the same as in the normal. The chromosomes of the ripening germ cells of this triploid female are irregularly distributed, and she is therefore semi-sterile; but amongst the few offspring produced when she is mated to a normal male there are some triploids, more diploids, and a few intersexes. The intersexes have three sets of autosomes (AAA) and two X-chromosomes (3A + 2X). Although such an intersex has the same number of X-chromosomes as has the normal female, it has more other chromosomes. This upsets the balance between the two, giving an individual that is somewhere between a normal male and female. If we suppose that the X-chromosomes contain more of the kinds of genes that stand for the femaleness, and the other chromosomes more for maleness, the intersex represents an intermediate balance. The protoplasm of the egg, which is the same in all these cases, plays no significant rôle in the result.

In the plant Datura it has been possible to obtain a complete series of types in each of which there is a third chromosome of one or of another pair. The additional chromosome leads to the formation of an individual that has recognizable differences from the original type, and from each of the other forms. Even two more chromosomes than normal may be added, which again gives a different type.

The question is sometimes discussed whether, in a species with many chromosomes, the upset of the balance by the addition of one more chromosome, is less injurious than in cases where the chromosomes are fewer. In general such a difference may be expected, but the individual differences of species must be taken into account before any such generalization can have much meaning.

The third problem involves the behavior of the chromosomes when they conjugate, i.e., pair off at the maturation stages. When the number is doubled, as in the tetraploid, there are four chromosomes of each kind. The usual rule is that they mate in pairs, which separate and divide as in the regular method of maturation. Each of the four resulting cells contains the double number, and after fertilization the tetraploid repeats itself. Such a tetraploid is self-perpetuating. A few cases are known in which a *hybrid* tetraploid has been produced by crossing two species; one set of chromosomes of one species and one of the other. For example: two species of Primula were crossed, each germ cell having nine (haploid) chromosomes; the hybrid had eighteen chromosomes. It was vigorous, but at the maturation of the germ cells the chromosomes derived from the two species failed to pair regularly, hence the ripe germ cells came to contain various numbers of chromosomes and were imperfect. The hybrid was therefore sterile. But a branch appeared on the hybrid with a double number of chromosomes. It arose, beyond doubt, by the failure of the protoplasm of a cell to divide after the chromosomes had divided. The result gave a tetraploid cell with two chromosomes of each kind from each species. From this cell the tetraploid branch arose. The branch produces flowers that are fertile, because the chromosomes can now mate in pairs, those of one species uniting with their like, and those of the other species also pairing. The offspring is a tetraploid hybrid, and the new species is self-perpetuating.

If a tetraploid is crossed to a diploid the offspring is triploid, which, as a rule, has imperfect germ cells. It is sterile, owing to the failure of all of the three like chromosomes to unite in pairs, which leads to irregularities in the distribution of the chromosomes when the germ cells mature. Occasionally, however, a germ cell may happen to get a haploid set of chromosomes. If two such unite the diploid condition is restored. The other possible combinations give unbalanced types of offspring.

Tetraploid cells, as described above, may appear on a plant by failure of the protoplasm to divide when the chromosomes separate. In fact, tetraploid cells are sometimes found in plant tissues. Such cells can be produced artificially by treating a cell in process of division with reagents that suppress the ensuing cell division. The daughter chromosomes then form a single nucleus with the double number of chromosomes, and this number remains throughout all of the later divisions. A germ cell may also rarely appear with the diploid instead of the haploid number of chromosomes. There can be no doubt that these also arise by the suppression of a protoplasmic division. Such tetraploid germ cells are larger than the normal ones.

The question naturally suggests itself as to whether, by artificially adding more protoplasm to a cell, it would remain larger after further divisions; and if so, whether the size of the chromosomes would remain the same. It has in fact been shown in certain cleavage stages that, after a cell has divided into a larger and a smaller one, the chromosomes in the latter are smaller than those in the foimer. Whether the genes are similarly affected is not known. At present the question cannot be answered, but if the amount of protoplasm could be increased artificially it might furnish evidence as to the presence in it of self-perpetuating units, independent of the genes. Since the genes are known to affect the amount of the protoplasm, as in tetraploids, it is evident that some interrelation does exist, and the experiment suggested above, if negative, might not be expected to throw further light on the question.

There is some evidence of the converse situation, when haploid individuals, or haploid cells in a diploid individual occur. Here it appears, in haploid plants, that the individual cells are smaller than those in diploids. In haploid males of animals, like rotifers, bugs and bees, the evidence at present is inconclusive. Occasionally haploid areas have been found in diploid Drosophila individuals. The cells in such areas are smaller. It is probable that in other cases the amount of protoplasm is correlated with the number of chromosomes, and it may seem that the protoplasm is dependent for its quantity on the genes, but still the question remains unanswered as to whether there are self-perpetuating units in the protoplasm that are inherited independently of the units in the chromosomes, or whether an increase or decrease in the amount of protoplasm is dependent on genic relations.

CHAPTER XI

PROTOPLASM AND GENES

The relative rôle of genes and protoplasm in the early and in the later development of the characters of the individual is one of the outstanding problems, both of genetics and of embryology.

The most significant evidence comes from the work on reciprocal hybrids of hermaphroditic plants. A cross made one way involves the egg protoplasm of one species; and made the other way, the egg protoplasm of the other species. Both hybrids contain the same genes—half from one parent and half from the other. Since in many cases the adult hybrids from the two kinds of crosses are identical, it follows in these cases that the protoplasm outside of the chromosomes is indifferent, at least when the adult stage is reached: i.e., the characters formed by the protoplasm are determined by the genes. The problem is, however, somewhat more involved, because in some cases there are known to be elements in the protoplasm of the egg that are self-perpetuating. For instance: there are plants, such as Pelargonium, that occasionally have white leaves owing to the absence of green chlorophyll bodies in them. These bodies are normally present as colorless plastids in the egg of a green plant, but not in the pollen tube—or if present are not carried with the nucleus into the egg. They are absent from the egg of the white part of white plants. Now, if pollen from the white branch fertilizes an egg of the green plant, a green plant results. Conversely, if pollen from a green plant fertilizes an egg of a white branch the resulting plant is white. The plastids which are the forerunners of the chlorophyll bodies are here transmitted only through the protoplasm of the egg.

An excellent illustration, showing partial dependence and partial independence of elements in the protoplasm to genetic constitution, has been described for reciprocal hybrids of two species of evening primrose, Oenothera muricata and Oe. suaveolens. The plastids of suaveolens cannot develop green chlorophyll in the genetic make-up of the hybrid, so that when suaveolens is the maternal parent the seedlings are white and die. In the reciprocal hybrid, with muricata as the mother, the plastids derived from the mother can develop green pigment in the genetic make-up of the hybrid, which is of course the same as in the reciprocal case. The young plants are green.

It is necessary to add a small reservation to the preceding state-ment. It appears that rarely a few plastids may enter the egg with the pollen nucleus. This complicates the results only to the extent that a few mosaics of green and white may arise owing to the irreg-ular sorting out of these plastids in the early divisions of the egg cells.

There may be, of course, other self-perpetuating bodies in the pro-toplasm, but as yet they are unknown, and even if they occur they are special cases, and do not seriously affect the general statement given above, based on reciprocal crosses.

There is supplementary evidence from experimental embryology, the earliest and clearest of which comes from crosses between differ-ent species of sea urchins. For example, the tempo of the first divi-sion of the egg is different in different species. When the egg of one species is fertilized by the sperm of another species, the rate of cleavage is that of the egg. The eggs of some species contain more pigment than the eggs of other species. The young hybrid larvae contain the same amount of pigment as the egg species. This is the pigment that comes directly from the unfertilized egg. In later stages, however, when new pigment develops, the effect of the sperm chromosomes may be seen. The number of the mesenchyme cells is different in different species. Thus Sphaerechinus has 38, Echinus 57. The hybrid from Sphaerechinus eggs has about 38 mesenchyme cells. One of the most striking examples of the relative influence of egg and

sperm is a cross between Cidaris and Lytechinus. The following
table gives, in hours, the time relations of different stages of development.

Cidaris	Hours	Lytechinus	Hours
Blastulae (swimming)	16 to 18	Blastulae (swimming)	5.5
Gastrulae (beginning)	20 to 23	Mesenchyme	8
Mesenchyme	23 to 26	Gastrulae (beginning)	9
Chromatophores	44	Chromatophores	15 to 16
Skeleton (beginning)	72 to 73	Skeleton (beginning)	15 to 16
Pluteus	120	Pluteus	24

When Cidaris eggs are fertilized by Lytechinus sperm the rate of
cleavage is that of Cidaris. The blastula also is like that of Cidaris.
In Cidaris the mesenchyme cells develop from the inner end of the
archenteron, while in Lytechinus they begin to migrate inward
before the gastrulation begins. The mesenchyme cells of the hybrid
arise from the sides and around the base of the archenteron, and not
from the inner end as in Cidaris. It appears, then, that the place of
origin is affected by the Lytechinus sperm.

The shape of the early triangular larva in the cross between
Sphaerechinus eggs and Strongylocentrotus sperm is that of the maternal type, but in the pluteus stage, that immediately follows, the
influence of the paternal chromosomes becomes apparent. Many
other descriptions have been given of older stages of hybrid plutei
between various species of sea urchins. In practically all of them
characteristics of both parent types can be detected. The few attempts that have been made to analyze the characters genetically—
such characters as the shape of the spicular skeleton, the form of the
larva, the kind of pigment, etc.—have not been very successful, for
while it is possible that one character is dominant over another, the
number of factors involved cannot be determined without rearing
another generation from the hybrids. Only in this way, as in most
genetic analyses, can the number of factors be determined by the
numerical relations of the types that reappear.

The general conclusions from the evidence are clear. The early stages of development are determined by the egg protoplasm; only later is the influence of the sperm chromosomes apparent. This means that the protoplasm of the egg has already been influenced by the genes of the egg itself. It takes time for the paternal genes to effect a change in the protoplasm of the cells of the embryo.

The hybridizing experiments are complicated by two other recognized conditions, namely, the influence of the medium and the irregular distribution of the chromosomes that sometimes takes place in the early cleavage stages of the hybrid.

The influence of temperature has been studied, both in the pure bred and hybrid plutei. Some influence is manifest. In the hybrid it appears that the greater or less "dominance" of the maternal or paternal influence is affected by temperature. The alkalinity or acidity of the sea water appears sometimes to affect the results.

It has been shown in certain crosses that elimination of a few or several chromosomes may take place either at the first or later divisions of the egg. Thus when Arbacia eggs are fertilized with Sphaerechinus sperm, about 18 of the chromosomes, presumably paternal, may be eliminated, which may account for the more maternal character of the hybrid pluteus. In the reciprocal cross no elimination takes place, yet the plutei are like those of the maternal race. The elimination of chromosomes upsetting the balanced effects of the genes may have the further effect of making the embryo pathological.

Teleostean fishes have also been used extensively in hybridization. Widely separated species can be crossed, but only in some of the crosses do embryos develop as far as the adult stage. The rate of cleavage is that of the egg species, except when delayed by abnormal separation of the chromosomes; but when closely related species are crossed it has been said that the tempo of cleavage may be hastened by the paternal genes if the male belongs to a type with quicker cleavages. The pigment cells that develop are not derived from the pigment of the egg, but pigment develops independently and may be influenced by the paternal genes. Elimination of chromosomes has been described in several combinations. Many embryos die in early stages, and this is probably due in some cases to elimination and irregular chromosome divisions; but in other cases is more probably due to incompatibility between the effects of the genes of the two species.

In the young embryo of certain races of the silkworm moth, pigment appears in the embryonic membrane from precursors in the egg. When races are crossed having differently colored embryonic membranes, the color is that of the egg race, i.e., it is maternal.

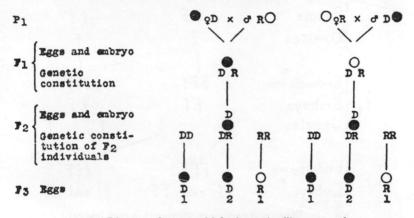

FIG. 76. Diagram of maternal inheritance in silkworm moth.

When adult hybrid moths are reared from these eggs it is found that their embryos develop the dominant color, irrespective of whether it has been brought in by the father or the mother, Fig. 76. That this result is really a case of gene inheritance is shown by breeding still another generation from the last. The embryos (F₃) from these are in the ratio of 3 to 1 (or, more accurately, 1:2:1). The result shows that the embryonic character is determined by the genes, which follow Mendel's law of segregation for a single factor pair, Fig. 76. In other words, maternal inheritance does not differ in principle from ordinary Mendelian inheritance. In this case the precursors of the pigment are already laid down in the egg under the action of the genes during the early formation of the egg, and are not changed at once by the genes of the sperm nucleus even though the latter are dominant. But in the latter case, the dominant genes do show their effect on the eggs of the next (F_2) generation. There are three kinds of genetic individuals in this generation, the dominant and recessive genes having segregated in F_1, and each acting on the development of the protoplasm of its own egg gives one or the other type of embryo in F_3.

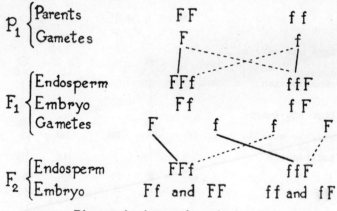

FIG. 77.   Diagram of endosperm formation in corn (maize).

There are some races of this moth that have two generations a year; others only one. It has been shown that this difference is also maternally inherited and is no doubt due to some peculiarity of the egg, or of the egg membranes, induced by the genes.

Similar effects are found in Indian corn in the formation of the endosperm of the seed. The endosperm is really a somatic tissue formed in the embryo sac—an enlarged cell—by the union of two maternal nuclei and one paternal nucleus. The color of the endosperm is determined by the dominant and recessive genes of these nuclei. For example: if a floury corn (ff) be used as the mother and a flint corn (FF) as the father, the endosperm is floury (ffF) which means that two doses or genes of floury dominate one of flinty, Fig. 77. If, on the other hand, a flint corn be used as the mother and a floury as the father, the endosperm (FFf) is flinty—two flinty genes dominating one floury. That this is the correct formulation is shown by breeding further generations, where the two factors of the embryo, F and f, segregate as a single pair. It will be noted in these two cases that the endosperm-protoplasm is not changed by the gene from the pollen grain because two recessives dominate.

Finally, an interesting case of maternal inheritance has been found in certain fresh-water snails. These snails, Lymnaea, are usually coiled in a right-handed or dextral spiral, but occasionally others with a left-handed or sinistrally coiled spiral are found. All the offspring of a given brood, including the hybrids, are all dextral or else all sinistral. It has also been shown that some sinistral mothers produce only sinistral broods, and that other sinistral mothers produce dextral broods. Conversely, some dextral mothers produce only sinistral broods and other dextral mothers may produce dextral broods. These facts were at first very puzzling from a genetic standpoint, but there is now a satisfactory explanation at hand. Suppose, as the evidence indicates, there is a dominant dextral and a recessive sinistral gene carried by a given pair of chromosomes. A self-fertilizing dextral snail that is heterozygous for these genes (Ll) produces after maturation two kinds of eggs, L and l. Similarly, there will be two kinds of sperm, namely L and l. Self-fertilization will give three genetic types of offspring—LL, Ll and ll; but all these individuals will be dextral because the cleavage pattern has been already determined in the egg by the dominant factor L before the polar bodies were given off. Of these three types the first two, LL and Ll, will produce only dextral offspring, but the other type ll, that has also a dextral shell, will produce only sinistral offspring. Since these snails may also cross-fertilize, provided dextral mates to dextral, and sinistral to sinistral, it is possible for the dextral female (arising as above) with the genetic constitution ll to mate with a dextral with the composition LL. All the offspring of

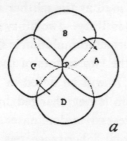

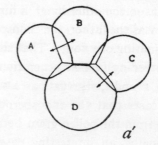

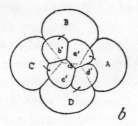

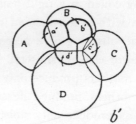

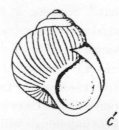

FIG. 78.   *a, b, c*, Left-handed type of cleavage and left-wound spiral shell of mollusc; *a', b', c'*, right-handed type of cleavage and right-wound spiral shell of mollusc. (After Conklin.)

such a somatically dextral female will be sinistral, since the undivided egg was under the influence of the two recessive genes (ll). These sinistral snails (Ll) in turn will produce only dextral offspring because the dominant factor L in the egg determines the type of cleavage of the eggs. It is evident, therefore, that dextrals of certain origins will produce only sinistral broods, and sinistrals of certain origins dextral broods. The heredity is Mendelian, but the appearance of the character is delayed for a generation. The result is unique, because the symmetry of the adult is determined, not by its own genetic constitution, but by that of the unreduced egg from which it arose. There is no contradiction in this to ordinary Mendelian inheritance of adult characters, if, as appears to be true, the symmetry is determined by the constitution of the egg before extrusion of the polar bodies, and, once determined, cannot later be re-

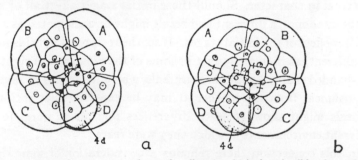

FIG. 79. *a*, Later cleavage stage of egg of mollusc in which the 4d cell has been given off on the right side; *b*, same with 4d on the left side.

versed, no matter what the genetic constitution of the zygote has become. That the egg is so constituted is shown by the early cleavage stages. At the four and eight-cell stages the spiral cleavages are in one direction in left-handed races, Fig. 78a, b, c, and in the opposite direction in right-handed races, Fig. 78a', b', c'. When the mesoderm cell, 4d, appears it is on one side of the prospective middle line in one type, and on the other side in the other type, Fig. 79a,b.

All the genetic evidence points plainly to the conclusion that the characters that develop in the protoplasm are ultimately traceable to the genes in the chromosomes. It shows, moreover, that the protoplasm of the egg formed under the influence of one set of

genes may after a time be changed over into protoplasm characteristic of another set of genes brought in by the sperm nucleus. It
shows, in the case of reciprocal hybrids, that the two sets of chromosomes derived from two different races produce the same end result
if sufficient time be allowed, no matter which protoplasm was present at the beginning.

On the other hand, there are a few cases where the reciprocal
hybrids are not identical. Aside from those cases where elimination
of chromosomes occurs in one cross and not in the other, differences
might sometimes be expected of another kind. For example: the
case of the snails shows that the very early stages of the embryo are
maternal in character. Should these initial stages affect all or some
of the later ones, a different end result might be anticipated as a sort
of after effect of the initial stages. If so, the reciprocal hybrids might
be different, as they are in fact in some of the crosses with right and
left-handed snails. Again: in mammals, where the foetus receives its
nourishment from the mother, it may be expected that reciprocal
hybrids might show at birth differences due, as it were, to the
different environments in which they were reared.

In this connection there remains a consideration of some significance. It might be claimed that only in those cases where the
protoplasms of the two parents are essentially alike is it possible to
obtain reciprocal hybrids that reach the adult stage; hence it is unsafe to argue from such cases to all cases. For, if there are really
differences in the protoplasm of different species that are independent of nuclear influences, in the sense that they can not be
converted one into the other by genic influence, then no late embryonic or adult stages would be expected. Therefore, until more evidence on a larger scale is obtainable, the central problems involved
cannot be regarded as settled for all cases.

The relative rôles of genes and protoplasm, exemplified in the
foregoing illustrations, bear on another question that has in
the past been discussed, namely, as to whether the protoplasm

determines the fundamental characters of the individual, and the genes only the more superficial ones. It need only be recalled that the experiments with left-handed and right-handed snails have shown that the genes can in time change one form of protoplasmic reaction into the other. If one pretends to be able to say what kind of character is fundamental and what superficial, surely the reversal of symmetry would seem to fall under the former class. The inheritance of the four blood groups in man, that follows the law for three allelomorphs, would also be regarded as a case where a fundamental character is determined by genic influence. Conversely, if pigment is regarded as a superficial character, then the case of the silkworm moth illustrates that this, too, may be determined by the protoplasm of the egg, but can be changed in later generations by the influence of the genes.

THE STABILITY OF DIFFERENTIATED TISSUES

For a long time it has been known that, under certain conditions, differentiated cells that have played a functional rôle in an adult individual may redifferentiate totally or partially, and later take a different part in the building up of a new individual. For example, at certain times of year the tissues of some ascidians may form a ball of materials in which all appearance of the old differentiation is lost, and later a new individual may develop. There is ro apparent continuity between the kind of cells of the old and of the new organs. The tissue of sponges and hydroids may be strained through bolting cloth, and, if allowed to settle on the bottom of a dish of sea water, several new individuals may develop by the coming together of the scattered cells. To what extent any cell may become a part of the new organism, or to what extent its former condition may influence its fate, is not entirely clear, but nevertheless there must be much redifferentiation in some of the cells.

There are some grafting experiments that, in part, give the reverse picture. For example: if a piece of the skin of the back of a newly hatched chick of a breed with marked feather characteristics is transplanted into a similar position on another chick of a different breed, Fig. 80, whose skin has been removed over the same area, the new feathers of the graft are like those of the breed from which the grafted piece came, Fig. 81. In the chick, only down is present at the time of operation, but the feather follicles have already been laid down. They give rise on the host, as stated above, to the same kind of feathers that they would have produced had they remained in place. Furthermore, if they are plucked out, or if they are molted,

FIG. 80.   Recently grafted chicks. (After Danforth.)

the new feathers that develop come from new follicles formed from the old ones and remain true to type. Here the genetic constitution of the individual cells determines the character of the feathers of the grafted skin, even when it comes to lie in an environment, the host, that has quite different feathers. There is no redifferentiation under these conditions.

On the other hand, it has been shown that other characteristics of the feathers are affected by the environment, or more specifically by the endocrines of the host. For instance: if the skin from the

rump of a young Plymouth Rock male is grafted on to a red female, the feathers are like those of the Plymouth Rock female in
this region and not like those of the male, Fig. 81a. In other words:
the endocrine secretion of the female determines certain characters

FIG. 81.   Adult fowls with grafts on rump. *a*, Red female with graft from Barred
Plymouth Rock male; *b*, Rhode Island Red male with graft from Barred Plymouth
Rock male; *c*, Buff Leghorn male with graft from Jersey Black Giant female; *d*, White
Leghorn female with graft from Rhode Island Red female. (After Danforth. *a* from
Journal of Heredity.)

of the new feathers. The rump feathers of a male fowl are long,
narrow and pointed; those of the female in the same region are
shorter and rounded.

The reciprocal graft shows similar effects. If, for instance, a piece
of the skin of the rump of a black female is grafted on the back of a

buff Leghorn male, the new black feathers are like the long, pointed feathers of the male, Fig. 81c. Here the absence of an endocrine present in the female accounts for the result, as seen when the ovary of a female is removed and her new rump feathers are like those of the male.

These results are consistent with the view that in such cases, when the protoplasm has once become differentiated, the end product is thereby determined—even when new cells are formed from the original ones that were set, so to speak, in this direction in an early embryonic stage. The results concerning the color characters of the feathers would seem to show that the genes have not been changed. On the other hand, the results concerning the shape of the feathers show that an internal environment, i.e., hormones produced in other parts of the body, also enter into some of the final stages of differentiation. The hormones may act first on the protoplasm of the follicle cells of the feathers, and, through the effects there produced, on the output of the gene; or they may act indirectly on the genes. There is at present no way of finding out which of these actions takes place. This evidence also is not decisive as to whether all the genes are acting, or some of them more than others.

Comparisons between the embryonic development of color patterns in spotted animals and the heredity of the same colors found in other races with uniform coat colors may be misleading. For instance: there are gray (agouti) guinea pigs, black guinea pigs, red guinea pigs and white ones. Single gene differences account for the inheritance of these colors. There are also other guinea pigs that are spotted. Some of these may have both gray and white areas of hair; others red and white; others black and white. At first sight it may seem paradoxical that a guinea pig that can develop areas of black hair should have white areas of hair if, as is the case, the cells of both areas carry all the genes. In those cases where the areas are definitely localized. i.e., distributed in definite regions of the body, their localization may not seem to be different from that of any other kind of somatic localization; but when, as in some spotted animals, the black and white spots are distributed at random, the paradox mentioned above is more apparent.

Genetic evidence has shown that, in mice, rats and guinea pigs there are definite genes for spotting. When two such recessive factors are present the animal is spotted, i.e., it has white areas and areas of some other color such as gray. The genes for gray are assumed to be present in all the cells; why, then, is not the hair gray everywhere? This question calls for further elaboration. For the production of pigment two substances at least must be present; if one of these is absent in a cell, or not produced there, no color will develop. In the presence of the genes for spotting, either the gene for gray fails in the white areas to produce the enzyme necessary for this color, or else the protoplasm of the cells in certain regions fails to respond to the output of such genes. Either alternative appears to land us on the horns of a dilemma, for, on the first assumption no explanation can be given as to what suppresses the gene; and on the second assumption no explanation is apparent as to why these protoplasmic differences are not found all over the body.

The same difficulty of explanation does not appear in animals that are pure white and still carry genes for gray or red or black. Here the recessive factor for white is supposed to be one whose normal partner when present produces one of the substances that is essential for the production of color. The recessive factor lacks this property—hence no color is possible.

The point to be emphasized here is that the spotting factor does not produce its effect in the same way as that for uniform white color, hence it is not legitimate to postulate that the enzyme relations are the same in both cases, in fact, they must be different. This, then, still leaves open the question as to how the spotting factor produces its localizing effects.

There are some skin grafting experiments showing that when

effects have once been produced the result cannot be reversed by changing the position of white and black areas in a guinea pig. When a piece of skin having black hair is grafted on a white area of the same guinea pig the hair remains black; conversely, white to black remains white. The same result holds for black and gray, or gray and red, etc. Obviously it is here no longer the location of the spot—whether white or black—that determines its color, for once determined it is fixed. This raises the question again as to whether the differentiation into black and white areas was really determined by localization in the embryo, or by some other kind of differential, such for example as a self-perpetuating protoplasmic element that gets irregularly distributed during early development. This leads to another situation that carries us back to the localizing problem.

For example: in an all gray guinea pig the hair on the back is quite different from that on the belly, both in color and length. If a piece of the latter is grafted on the back it retains all the characteristics of belly hair. If, as one would ordinarily assume, the difference in the back and belly is a problem of embryonic localization, why should we refrain from using the same argument for the spotted animals, even although the irregularity in the distribution of spots makes the problem seem to be somewhat more difficult?

CHAPTER XII

LARVAL AND FOETAL TYPES

In all of the great groups of animals whose young develop in water, especially those in salt water, there are free-swimming larvae typical for each group. These larvae distribute the species widely. They must obtain food to complete their development, since they come from small eggs not containing enough yolk to carry them through to the adult form. The typical larvae of each group have been given a special name; for, while the larvae of each species show characteristic features, yet within the group they have the same fundamental structure. A few of the better known types may be described, and the interpretation of their presence discussed.

In the group of echinoderms, the sea urchins pass through a pluteus stage; while other subdivisions, starfish, crinoids, and holothurians also have somewhat similar and characteristic larvae. The pluteus, Fig. 82, is a small transparent larva, with a calcareous skeleton and with a somewhat elaborate band of ciliated cells over the surface, by which it swims actively about in the sea. On one side there is a large mouth leading to a tubular oesophagus which opens into a big stomach. This is followed by a short intestine which opens on the dorsal side. Between the digestive tract and the outer wall there is a fluid space containing loose mesenchyme cells, from some of which the calcareous skeleton has been formed. In addition there is a small sac on the dorsal side that opens by a small tube to the exterior. Its walls come later to form the lining of the body cavity and water vascular system. When fully formed a sudden change takes place in the larva. Around the mouth a mass of cells appears derived from the larval organs, which mold themselves into the very different form of the young sea urchin or starfish. The old skin is slowly absorbed into the growing urchin and the larval skeleton disappears. A sudden and great change like this is called a metamorphosis. The embryo sinks to the bottom, its movements are now brought about by a system of tubular feet extending along the five rays of the newly formed body. The larva was bilaterally symmetrical, the adult is radially symmetrical.

The marine annelid worms and molluscs pass through a free-

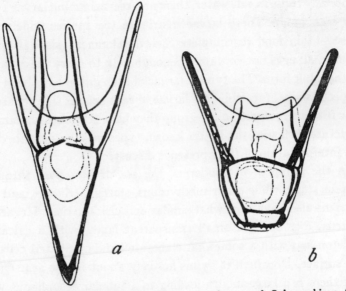

FIG. 82. *a*, Pluteus of Strongylocentrotus; *b*, pluteus of Sphaerechinus. (After Herbst.)

swimming stage known as the trochophore, Fig. 83a. It has two bands of cilia and at its anterior end there is a tuft of cilia with a very simple brain beneath it. The digestive tract consists of mouth, oesophagus, stomach, intestine, and anus. Between its outer walls and the inner body wall is a large space with a few muscle cells, running around and along the walls. In the annelid there is a small group of mesoderm cells at the posterior end from which the future segments and the muscles will develop. The transformation of the trochophore into the worm is a gradual process. The posterior end elongates, Fig. 83b; the mesoderm forms paired blocks of cells, one pair after another. Each pair contains a cavity; its inner walls form the muscles of the digestive tract; its outer those of the body wall. Between consecutive segments septa are formed; the head end grows

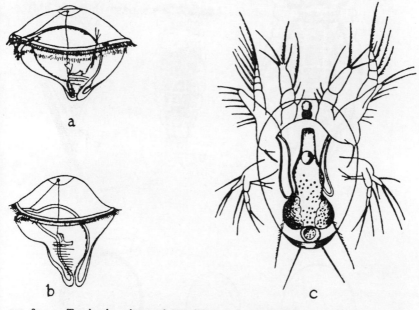

FIG. 83. *a*, Trochophore larva of Annelid; *b*, later stage of same, showing elongation of posterior end and development of segments; *c*, Nauplius larva of crustacean. (After Hatschek, from Korschelt and Heider.)

smaller; the brain enlarges, and along the ventral side an ectodermal thickening becomes the nerve cord.

The transformation of the trochophore of the molluscs is somewhat different. A shell develops in the dorsal side, as a secretion of the skin, that later comes to surround the animal when it contracts. A thickening of the ventral walls forms the muscular foot. The adult is not segmented as is the annelid, but consists of a single anterior segment, or at least not more than one additional one. The adult mollusc and annelid have evolved along quite different lines.

The presence of a common larval type in both groups has been

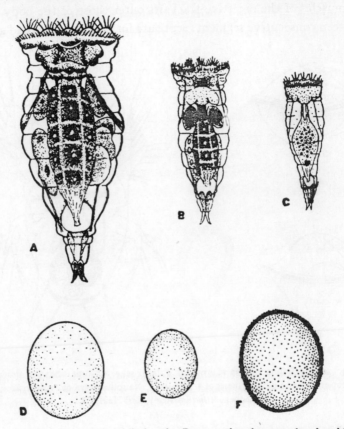

FIG. 84.   Hydatina senta. *A*, adult female; *B*, young female soon after hatching; *C*, adult male; *D*, parthenogenetic egg; *E*, male-producing egg; *F*, resting egg. (After Whitney.)

interpreted to mean that it is the ancestral form from which they have evolved. It seems more probable, when other considerations are taken into account, that the presence of the trochophore in both groups indicates rather that these groups have come from a common ancestral group that passed through a trochophore stage which is still retained. The identity of their cleavage type supports this view, since the cleavage stages were certainly not ancestral adults.

There is a group of fresh and salt-water animals, the rotifers, Fig. 84A, that resembles the larval trochophore, Fig. 83, in many respects. The rotifers never pass beyond this stage, but develop a pair of ovaries or testes. It has been suggested that the rotifers are the arrested trochophore stage of some ancient annelid (or mollusc) that no longer reaches the former adult condition.

Most of the lower crustacea have a small free-swimming, six-legged larva, Fig. 83c, called a nauplius, which at one time was also regarded as the ancestral type from which all the crustacea have evolved. There are, however, better reasons for regarding it as the conservative larval type of the ancestors of the crustacea that has been retained.

Most of the lower insects pass through a gradual series of stages, each more advanced than the last. Each stage ends with a molt, when the outer cuticle, that has stopped further growth, is thrown off. A new soft skin has meanwhile developed under the old cuticle. A marked increase in size immediately takes place before the new cuticle hardens. In the highest groups of insects—beetles, butterflies, flies, bees, and ants, there is a characteristic larval stage very different from the adult. In this stage, the organism is called a maggot or a caterpillar. The transition from larva to adult in these groups is abrupt. A true metamorphosis takes place. A quiescent stage intervenes, known as the pupa, in which, from discs of undifferentiated cells, the primordial discs, the organs of the adult are formed inside the pupal case. Most of the larval organs are totally lost and absorbed.

Without attempting to describe this metamorphosis of insects, it will suffice to point out that the larval types of these higher groups cannot be regarded strictly as the ancestral larvae, although some of the features of such larvae may have been utilized. The larvae are

themselves highly adapted forms that have departed from the original type, perhaps as much so as the adult itself from other members
of the ancestral group. For example: the caterpillar of the moth or
butterfly chews its food by means of horny jaws, while butterflies
have a tube to suck up the nectar of plants. The caterpillar has
evolved a new set of legs. The compound eyes are absent. Larva
and adult have become so specialized that a gradual transition of
one into the other is impossible. Hence, when the caterpillar has
reached its final stage it ceases to move, spins a cocoon about itself,
and from its primordial discs of undifferentiated cells the organs of
the butterfly develop. Just how this adjustment took place histori-
cally is not easy to say, although there are transitional stages in
some of the other groups that make the difficulties appear not in-
surmountable.

In the vertebrates there is no very characteristic larval stage,
although the tadpole of the frog represents such a type for the amphibia. It, too, like the caterpillar, is a specially adapted larva, and
undergoes a sudden metamorphosis at the end of the larval period.

The tadpole resembles in many particulars the corresponding
stage of the simpler group of amphibians—the newts and salamanders—and since the larvae of the latter are, except for size, very like
the adult stages, it may appear that it is the adult salamander that
is repeated in the development of the frog. This is, in fact, the implication of Haeckel's so-called biogenetic law. There are, however,
many reasons for regarding the latter interpretation as erroneous.
The tadpole is not the recapitulated stage of the salamander, but a
stage common to the frog and salamander during their development.
When the tadpole has grown to full size—which may take a few
weeks or may take one or more years in different species—it ceases to
feed, and in the course of a few days it undergoes retrogressive
changes in some of its organs and progressive changes in others. The
tail shortens and its internal tissues are absorbed. They are eaten up
by the phagocytes, or white blood cells, which carry the products of

digestion back into the body. Much of the gill region is absorbed. The digestive tract of the tadpole is a long coiled tube. At the time of metamorphosis it shortens, and its walls thicken. The horny teeth in the mouth are lost, and the mouth widens. The gill slits close up, and the gills are absorbed. The blood vessels that run through the gill arches are in part absorbed, and in part transformed into certain of the adult vessels. The lungs, that have meanwhile developed as a pair of sacs, growing out of the œsophagus, begin to function as air is taken in through the nose. Further conspicuous progressive changes take place. The fore and the hind legs, that are present in the tadpole as small rudimentary organs, suddenly begin to enlarge. The fore legs are very small until metamorphosis begins, when they grow rapidly and break through the opercular fold that covers them until this time. The hind legs, that are longer, also grow rapidly. It is interesting to note that two processes take place simultaneously, which are the reverse of each other; some organs degenerate, while others grow.

The metamorphosis is intimately connected with changes in the thyroid gland, which is an organ of internal secretion producing a hormone, i.e., a substance (thyroxin) that is absorbed by the blood from the gland, and initiates some of the changes in the body described above. The transition from tadpole to frog can, in fact, be brought about long before the change is due, by keeping the young tadpole in water containing iodine compounds, or by injecting into it similar substances.

One of the salamanders, the axolotl of the Mexican lakes, reaches sexual maturity when the gill slits and gills are still present. It breeds usually in this condition. Its adult form—a land salamander —develops only under exceptional conditions. The ripening and functioning of the reproductive cells, eggs and sperm, while the individual is still in the larval stage, is called neoteny. Its occurrence is suspected in other cases amongst the long-tailed amphibia, and in several other groups of animals. For example: the ascidians pass through a free-swimming larval stage, Fig. 85a, that resembles the tadpole of the frog. There is, also, a group of free-swimming pelagic ascidians, the Appendicularia, Fig. 85c, that have the essential structure of the tadpole stage of other ascidians. The appendicularians reproduce, and never change into the form characteristic of the other members of the group They are probably a neotenic type.

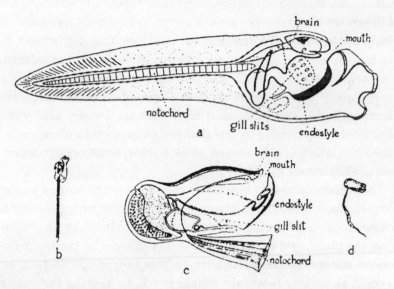

FIG. 85. *a, b*, Tadpole stage of Ascidian; *c, d*, free-swimming appendicularian.

In contrast to those animals that posses many small eggs giving rise to free-living larvae, there are other animals that produce a few large eggs containing enough food to carry the embryo through its later development to a stage that resembles the adult except in size. The development may be said to be direct. The term foetus is used for embryos with direct development carried to a relatively late stage within the mother, to whose uterus the foetus is attached by a special embryonic organ, the placenta, bringing its circulation into such a relation with that of the mother that nourishment is absorbed from her.

In most of the great groups there are species having a direct development, and these usually have very large eggs. The "yolk" of the hen's egg has twenty-seven million times the bulk of that of amphioxus. The best known and most interesting cases are those present in the vertebrates. The lizard and the bird have enormous eggs. The young when hatched, or born, have the form of the adult. In the mammal the development is closely similar to that of the bird and the lizard despite the fact that it begins as a very small egg.

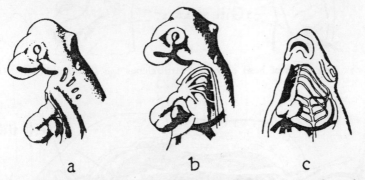

a   b   c

FIG. 86.   *a*, Head of chick in side view, showing four gill slits; *b*, aortic arches; *c*, heart and aortic arches of fish. (After Hesse.)

A comparison of the development of these types with one another, and with lower members of the group, brings out some interesting relations. First, the presence of organ systems such as the gill slits of lizards, birds, and mammals, Figs. 86a, b, that are reminiscent of the gill slits of the free-living larvae of fish, Fig. 87, and amphibia; second, the presence of embryonic novelties that adapt the embryo to new conditions, such as the amnion and allantois, Fig. 31. In addition there are many minor features that come under one or the other head. The comparison goes back to the developmental stages of fish and amphibians. The fishes, both the cartilaginous group (sharks), and the bony fishes (teleosts) breathe by means of a gill-system, Figs. 86c and 87. This consists of gill slits on the sides of the neck that open on the inside into the oesophagus and on the outside into the water. Between the gill slits are gill arches which serve to carry the blood from the ventral heart to the dorsal aorta. On the

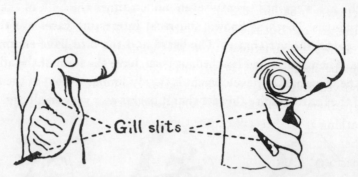

FIG. 87. Side views of head of young cartilaginous fish showing gill slits. (After Sedgwick.)

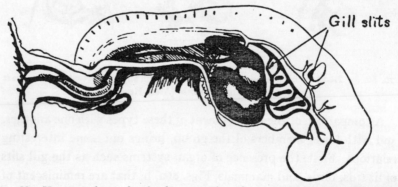

FIG. 88. Human embryo, showing heart, aortic arches and gill slits. (After His, irom Marshall.)

arches are many filamentous gills carrying capillary blood vessels. The blood is aërated in the gills. Breathing consists in taking water into the mouth and passing it through the gill slits out over the gills. This system appears very early in the development of fishes, irrespective of whether they come from large eggs, as in sharks, or from small eggs, as in the bony fishes. The system does not function, however, until the young hatch.

In lizards, birds, and mammals, Fig. 88, gill slits also appear in the embryo at a stage corresponding to that of fishes. They never function as breathing organs, but soon close up. Gills do not appear on the arches. But the blood vessels that carry the blood through the arches are functional from the beginning, carrying the blood from the heart to the dorsal aorta. In later stages some of these blood vessels persist, to become parts of the adult circulatory system; others disappear. The significant points are, first, that the breathing apparatus appears at the same stage in the embryo of the higher group as in that of the lower; second, that it has a function to perform in the embryo, carrying the blood from the heart to the dorsal side; and third, that the aortic arches represent a structural stage from which the more specialized system of the adult emerges.

The amphibia supply an interesting intermediate condition. They are nearer to the higher types than are the fishes, even although they may not be in the line of immediate descent. Moreover, some of the amphibia represent one of the earliest types of vertebrate animals that left the water to live on the land, and show a transition from gills to lungs. Those that now live on land, the frogs and many salamanders, resort to water to lay their eggs, and in the water the young stages are passed through. These young respire first by means of embryonic external gills, which are succeeded by internal gills, like those of fishes. The gill system is absorbed when the older larvae leave the water to live on land, and its absorption and transformation are essentially the same as in birds and mammals. Thus the young frog passes through its early stages in the same way as do water-breathing fish, and its adult stage is like that of air-breathing reptiles, birds, and mammals.

The similarity among the young stages of the higher groups of vertebrates was noted and commented on more than a hundred years ago, at a time when the theory of organic evolution was not in vogue. The resemblance was given only a formal explanation. The most famous statement is known today as von Baer's laws, stated in 1828. His statements are often confused with Haeckel s biogenetic law, which is, however, entirely different, and rests on the acceptance of the descent theory. Von Baer's four laws were as follows:

1. The most general features of a great group appear earlier in the embryo than do the special features.

2. Out of the most general, the less general structures arise, and so on until at last the most special structures appear.

3. Each embryo of a given adult-form of animal, instead of passing through other adult forms, on the contrary diverges from them.

4. Fundamentally the embryo of a higher animal is never like the adult of another animal form, but only like its embryo.

The biogenetic law, so called, is more explicit. It states that the embryos of members of the higher groups pass through the adult stages of lower members of the same group, which have, as it were, been condensed into stages of development. This statement represented so dramatically certain features of development that it became one of the most popular theories of biology. Its very simplicity conceals its implications and gives the appearance of an "explanation" when it merely serves to hide a historical relationship. Now that we have a better understanding of how new genetic variations are introduced, we find that there is no fixed relation between the original character and the appearance of a new one as implied in the biogenetic "law." In many cases, perhaps in most, a new end character simply replaces the original one. The embryo does not pass through the last stage of the original character and then develop the new one—although this may happen at times— but the new character takes the place of the original one. For example: when the mutant type of Drosophila with rudimentary wings develops it does not pass through a stage with large wing pads or large primordial discs, but develops directly from discs which are proportionately reduced in size. Similarly for almost all of the known mutant types; and this is true not only in cases where the new organ is reduced, as in vestigial wings, but also where the new character may add something more than was there before.

It is true that there are a few cases where the new character is superimposed on the old one, and the latter may be to some extent retained. For example: several beetles have black spots on a yellow background. Some mutants are known that are black. These pass through a stage when the spots are present on a light background, but the latter soon becomes black, obscuring the spots. Here one may say that a new character is added to the old one at the end of the series, and in such cases it might be legitimate to state that an ancestral adult stage is retained in the development of the later stage, but these cases are the exceptions. The rule is as stated above, that the original end stage is dropped out and replaced by a new character.

While Haeckel held that embryonic stages represent for the most part ancestral adult types, he was familiar with embryonic adaptations that could never have been adult stages, such as the allantois, amnion, and yolk sac of birds and mammals. He spoke of these as falsifications of phylogenetic ancestral stages. They are embryonic adaptations. The admission was soon found to open the door to skepticism with respect to many larval and embryonic stages that had been interpreted as ancestral, such, for example, as the nauplius of crustacea, and the trochophore of annelids, which are now regarded only as specialized larval forms whose widespread occurrence in a group is not sufficient evidence that they are ancestral adult stages. In other words: if the recapitulation theory is a "law," it has so many exceptions that it has become useless and often misleading. Moreover, as stated above, it carries an implication with respect to the way in which new characters are "added" that is inconsistent with a large body of definite information.

The theory of the gene with its implications is entirely in accord with what is known concerning the introduction of new characters into the development. A mutational change in a gene may be such that its effects become apparent only when the last stage of development is reached, which may alter that stage completely. Again, the genic change may be of such a sort as to affect the larval stage, altering one or more of its characters. Such a change might or might not affect the adult stages. It would seem a priori that the earlier the change is brought about in a conventional type of development, the more likely will it affect also later stages. So complicated is the series of events of development that it may be much more difficult to introduce a change at an early stage without bringing disaster to the later stages than the reverse. A plausible case might be made out on these grounds for the conservation of the early embryonic and larval stages of the great groups.

Neither all of von Baer's laws nor Haeckel's biogenetic law are in accordance with the fact of variation and development known at the present time. A much simpler and more consistent explanation is possible. The gill slits of the mammal are not to be compared with those of an adult fish, but with those of a young fish of the same stage. As the higher vertebrates have evolved from the lower ones they have retained many features of the early stages of their ancestors. If anyone wished to do so, he might say teleologically that this is the only method "known" to them by which to reach their adult stages; or, put in another way, they have found that the simplest manner in which to develop their more elaborate structures is to build on those that have been used for millions of years. This might be called the principle of embryonic conservatism. Stated in a less anthropomorphic way this principle, which need not be given the high sounding status of a law but which may be called the principle of embryonic conservatism, is that the higher and lower members of groups pass through the same stages of development.

CHAPTER XIII

PARTHENOGENESIS

The history of parthenogenesis is one of the most interesting chapters of biology. Aristotle had a suspicion that some of the castes of the honey bee reproduce without mating, but his picture was wrong in most respects. It was not until the end of the seventeenth and the beginning of the eighteenth century that a few observations were made indicating that the eggs of some moths may develop when the female has not mated; but it was later in the middle of the eighteenth century (1737) that convincing evidence of parthenogenesis was reported by Reaumur, whose observations led Bonnet, a little later, to undertake a classical series of observations on the reproduction of aphids. Bonnet's writings and speculations aroused wide interest in parthenogenesis—an interest that extends to the present day.

The demonstration of parthenogenesis in bees by Dzierzon (1845) started new interest and much discussion. An extensive literature grew up, including observations on other insects. The extension of Dzierzon's theory and its application to other animals, notably by Von Siebold and by Leuckart, carried the problem into wider fields of biology, especially later in relation to the number of chromosomes in the eggs of those animals that develop by natural parthenogenesis. Later still the experimental work on eggs, incited to parthenogenetic development by artificial, i.e., unnatural agents, suggested other problems concerning the rôle of the spermatozoön in fertilization.

It was largely through Weismann's observations (1886) and theories that a significant relation between chromosomes and parthenogenetic development was brought into the foreground. He pointed out that eggs of certain animals that normally develop without fertilization, give off one polar body only, while the fertilized eggs of the same animals give off two polar bodies. The observation had been made a few years earlier, but its significance not realized. It is now found to hold, not only in certain crustacea (cladocerans and ostracods) but also in rotifers, bugs, bees and ants.

It was soon recognized that in many of the cases of natural parthenogenesis the unfertilized eggs produce only males; in other cases only females. The production of males only from unfertilized eggs is of widespread occurrence in many species of bees and related forms. In other groups where an alternation of generations (a bisexual generation alternating with a parthenogenetic generation) occurs the situation is more complex. An excellent recent summary of the whole subject is given by Vandel. A few examples will serve to illustrate the principal facts that are known.

NATURAL PARTHENOGENESIS

In the rotifers, especially in Hydatina senta, not only has the relation of the chromosomes to the sexual and parthenogenetic phases of the life cycle been made out, but there is much experimental work bearing on the causes that lead to the transition from one phase to the other. The life cycle consists of a line of females whose eggs develop parthenogenetically (virginally), which are succeeded by sexual males and females, Fig. 89 and Fig. 84. The eggs of the parthenogenetic females give off one polar body. At this time each chromosome splits longitudinally, as in the somatic division; one daughter half goes into the polar body, the other half remains in the egg, which then has its full complement of chromosomes. This method of reproduction continues through several or many successive generations of females—in fact, indefinitely if the external conditions are unchanged. After a time a female may appear that produces eggs of a different sort. This female is called the male-producing female; her eggs have two possible fates: First, if the female is not impregnated by sperm from a male, she produces small eggs that give off two polar bodies; these eggs develop into haploid males. The chromosomes have been reduced in number at one of the two divisions of the polar spindles, the egg retaining the half number only, and this number is characteristic also of the cells of the male.

HYDATINA SENTA

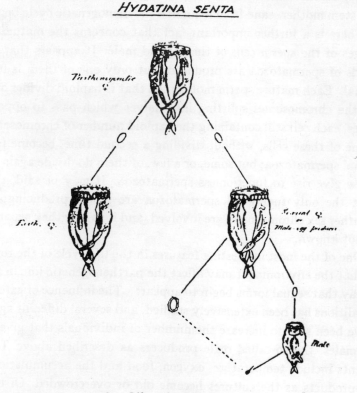

FIG. 89.   Life cycle of Hydatina senta.

The male itself is a very rudimentary affair that can not feed. Its interior consists largely of sperm cells. Second, if this same female has been impregnated with sperm from a male soon after she emerges from the egg membrane, the spermatozoa find their way to the young egg cells, and one enters each. These eggs then grow to a larger size than do the unfertilized male eggs, and become covered with a shell. Each egg gives off two polar bodies, and at one of the two polar divisions the chromosomes are reduced to the half number. The egg nucleus and the sperm nucleus then come together to form the segmentation nucleus, which consequently has the diploid number of chromosomes. This embryo always develops into a female— the stem mother—and from her a new parthenogenetic cycle begins.

There is a further important fact that concerns the maturation stages of the sperm cells of the haploid male. It appears that two kinds of spermatozoa are produced, but only one of them is functional. Each mature sperm mother cell that is haploid divides once, all the chromosomes splitting into halves which pass to opposite poles, each cell still containing the haploid number of chromosomes. Some of these cells, without dividing a second time, become functional spermatozoa; but some, or a few, of them do divide again and these give rise to functionless spermatozoa. It may be said, then, that the only functional spermatozoa are female-producing; but whether sex chromosomes are involved, and if so how they separate, is not known.

One of the most interesting features in the life cycle of the rotifer is that the environment may affect the parthenogenetic line in such a way that sexual forms begin to appear. The influence of external conditions has been extensively studied, and several different agents have been said to increase the number of individuals that give rise to males, the so-called male producers as described above. These agents include temperature, oxygen, food and the accumulation of by-products as the cultures become old or overcrowded Of these only one, namely food, has been convincingly shown to be effective. For instance, if Hydatina is fed on a colorless flagellate protozoön (Polytoma) it continues indefinitely as a parthenogenetic line; but if at any time a green flagellate, Chlamydomonas, is introduced in abundance, nearly all of the next generation of females produce male eggs, or, if they are fertilized, sexual eggs. The ending of the parthenogenetic line and the beginning of the sexual phase is caused by a

change of food. As stated above, the same individual that produces
male eggs also produces sexual eggs if early fertilization occurs. The
environment may be said to be sex-producing but not sex-determining. The presence or absence of a spermatozoön is sex-
determining.

One group of crustaceans, the cladocerans, show certain similarities to the rotifers in that there may be a long line of partheno-

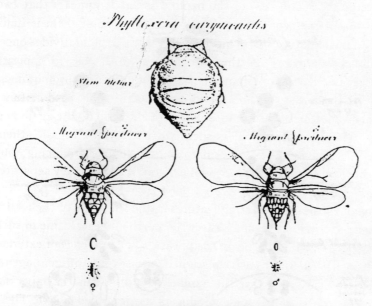

FIG. 90.   Life cycle of Phylloxera caryæcaulis.

genetic females that are diploid and set free only one polar body. As
a result of a change in external conditions females may appear that
give rise to males or to sexual eggs, but the conditions that determine whether an egg is of one or the other kind are apparently
not the same conditions as those in rotifers. For example: in the
cladocerans the male-producing females appear when the cultures
become crowded, but in rotifers crowding suppresses the appearance
of such females. In cladocerans it has been suggested that the

accumulation of excretory products of the animals themselves (or of the bacteria present) brings about increased male production. The addition of uric acid, or chloretone, or carbon dioxide increases the

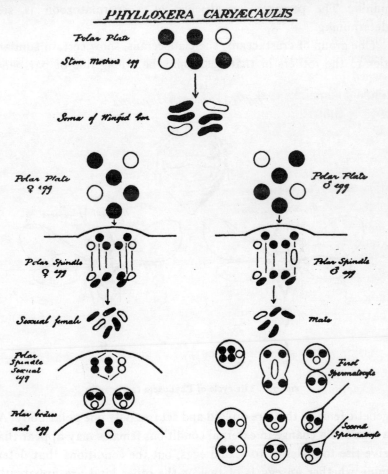

PHYLLOXERA CARYÆCAULIS

FIG. 91. Chromosomal cycle of P. caryæcaulis.

percentage of males. In the rotifers there is no evidence that these agents cause the change; on the contrary, old cultures of horse manure reduce the percentage of male-producing daughters.

In insects. the aphids and phylloxerans, Fig. 90, also have an alternation of parthenogenetic and sexual generations. In the former it has been shown that the length of day, i.e., of light hours, is one of the agents bringing on the change, but whether directly or in some indirect way is not clear. There are two further facts known in these two groups that explain certain relations not established for the rotifers: (1) The parthenogenetic females have the diploid number of chromosomes and give off one polar body, Fig. 91. At the end of the parthenogenetic line a female appears whose eggs also give off only one polar body, but when all the other chromosomes divide one of the sex chromosomes passes entirely into the polar body leaving behind its mate, the other sex chromosome, Fig. 91 (to the right). This egg then gives rise to a male, that has the diploid number less one X. (2) When the sperm cells in this male undergo the first maturation division each of the chromosomes conjugates with its mate except the X, that has no mate. The members of each pair then move to opposite poles and the single X moves toward one of the poles, Fig. 91. The cell getting the X becomes larger than the other one and the two daughter cells separate. One of them has a haploid set (including the X), the other has a haploid set less the X. The larger cell then undergoes the second maturation division, each chromosome, as well as the X, splitting. Each becomes a functional spermatozoön, female-producing, since it carries an X-chromosome. The smaller cell, the product of the first division, then degenerates.

In other females appearing at the same time, or a generation later than the male-producing female, sexual eggs develop. They give off two polar bodies, and retain the haploid number of chromosomes including one X, Fig. 91 (to the left). Fertilized by one of the female-producing sperm of the male, the full number of chromosomes (2A + 2X) is restored. This egg, called the resting egg, produces the stem mother, the progenitor of a new line of parthenogenetic females.

Here again the sex chromosomes play a regular rôle in sex determination, while the environment changes parthenogenetic reproduction into the sexual phase.

There are several species of animals that are represented by females only in certain localities, and by males and females in other localities. The former reproduce by parthenogenesis; the latter reproduce by fertilized eggs. For example: one of the isopod crustaceans, Trichoniscus, is represented in the north of Europe only by females; these are triploids with 24 chromosomes; in the south, males and sexual females are found which are diploid, with 16 chromosomes. If the eggs of these sexual females are not fertilized they fail to develop. There are here, then, two distinct races of the same species, and since they cannot cross they may be said to be two species which, except for their mode of reproduction, are identical.

In another crustacean, Artemia salina, there are parthenogenetic races that are tetraploid with 84 chromosomes, and sexual races that are diploid with 42 chromosomes. There are also cases known amongst the grasshoppers where, if the eggs are fertilized, they give rise to males and females—sex being regulated by the two kinds of spermatozoa; but if the eggs are not fertilized they produce females. Certain species of phasmids reproduce solely by parthenogenesis; other species are sexual. Many other instances could be given where parthenogenesis plays a rôle in the life cycle.

These examples, that might be multiplied, serve to show that parthenogenesis is widespread in the animal kingdom. It is also known in many plants. These cases emphasize the fact that eggs in themselves have the power to develop. They give a different picture from that usually presented, which lays all the emphasis on the initiation of development as a result of fertilization. It is better to say, I think, that the chief function of the entering sperm is to remove a block that holds the egg in check. From this point of view the parthenogenetic egg differs from the ordinary egg in that the block is absent. Of course other situations are also involved, especially the number of chromosomes that the egg comes to contain, the number of polar bodies it gives off, and the rôle of the sperm in introducing a division center into the egg, or causing one to develop there. The situation will be more obvious when the behavior of eggs artificially incited to develop has been described.

ARTIFICIAL FERTILIZATION

As early as 1866 it had been found by Tichomiroff that unfertilized eggs of the silkworm moth, if rubbed between two pieces of cloth or immersed for a brief time in sulphuric acid, will begin to develop (6 percent) and form embryos. The experiment has been repeated more recently in several ways, and on the whole confirmed, but subject to the reservation that even without treatment the unfertilized eggs of certain races of this moth may also begin to develop. About thirty years later it was found that the unfertilized eggs of a sea urchin and an annelid could be incited to start development, at least as far as the cleavage, by adding certain salts to sea water. It was not until 1900 that Loeb succeeded in causing the unfertilized eggs of the sea urchin to develop into swimming embryos by changing the chemical constitution of the sea water. Since that time it has been shown that eggs may be incited to develop by many different kinds of agents. The earlier expectation that "chemical fertilization" would elucidate normal fertilization cannot be said to have been realized, but a consideration of the possible changes induced in the egg by artificial reagents has led to some interesting speculations concerning the nature of the changes that the entrance of the spermatozoön may bring about. The difficulties in interpreting the problem are due not to lack of evidence, but to the discovery of so many ways in which the development may be started. To point out any one property common to these reagents has not been possible. It is the change in the egg itself that has so far baffled all attempts at an explanation. This will be apparent when some of the methods employed have been reviewed.

In one of the earliest procedures sea urchin eggs were left for about an hour in sea water to which magnesium chloride was added (50 cc. sea water plus 50 cc. of $2\frac{1}{2}$ normal $MgCl_2$) and then returned to sea water. An improved method consisted in leaving the egg for two hours in hypertonic sea water, then in sea water to which a little ethyl acetate had been added. On returning the eggs to sea water many developed in a normal manner. Certain acids also gave similar results. The first indication that development will take place is found in the lifting of the membrane. This occurs also normally when the sperm touches the egg. A transparent membrane appears that lifts off the egg and rather quickly draws away from the surface leaving a fluid space inside. The formation of the membrane was supposed to be one of the essential factors in starting the egg, but other methods have since been found which suppress membrane lifting yet do not interfere with normal development.

The eggs of starfish can be started by treating them at a particular moment with sea water saturated with carbon dioxide. Even shaking the eggs will start the cleavage, and also treatment with butyric acid. The egg of the frog may be incited to develop by pricking it with a fine needle. Whether or not the presence of lymph or blood on the needle is essential for success is not certain, but some experiments made to test this point seemed to show that in the absence of lymph the development may begin but a division figure does not develop, while when lymph is present a few eggs may divide. At best only one egg in thousands punctured may give rise to a young embryo, and of these it is very rare for one to develop as far as the frog stage. The eggs of several kinds of annelid worms can be started with potassium and sodium chloride, others with acids and by heat, and others by diluting the sea water.

It is obvious from the evidence that, while there are specific agents that give the best results for each kind of egg, these agents differ so widely that it is not possible to make a general statement as to the way in which they act. Loeb laid great emphasis on the cytolytic effect of these agents on the surface layer of the egg. In a sense this means a destructive effect. He supposed that the double treatment gave better results, because the second treatment counteracted the effect of the cytolytic action of the first. But since a single treatment suffices in many cases it is not obvious that the effect is cytolytic, and even if so the nature of such a change is still not apparent. What needs to be found out is how the reagents remove the block

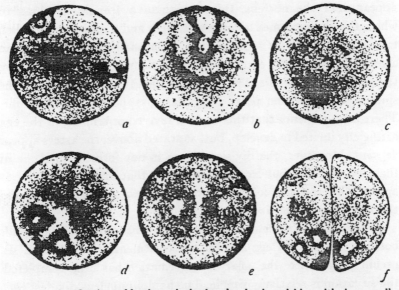

FIG. 92. *a, b, c,* Sections of frog's egg incited to develop by pricking with clean needle, showing the development of the aster near the path of puncture; *d, e, f,* sections of eggs pricked with needle in presence of blood (double treatment). (After Herlant.)

that inhibits the changes that the egg itself is prepared to undergo. That some of the reagents employed do act injuriously on the egg can not be doubted, since, if left longer than the optimum time in these fluids the eggs are injured beyond recovery. The initial change in the surface may be a step in that same direction, but just what this involves is not sufficiently explained by calling it a cytolytic process—unless this means any kind of change that injures some structure of the surface or that holds the development in check.

In more recent years a great deal of attention has been paid to the changes that can be seen in the egg itself after treatment with activating reagents. These involve in the main the development of the division centers in the protoplasm, and the behavior of the nucleus of the egg. Briefly, it has been found that all successful procedures lead to the development of asters in the egg, and the best treatments are those in which only a few, preferably two asters appear near the nucleus. Between them a spindle develops to which the chromosomes become attached, and a division may then take place. It will be recalled that in normally fertilized eggs an aster appears near the entering sperm nucleus which divides to form the poles of the spindle. The aster that lay at the inner end of the spindle of the second polar body has disappeared by this time. There is no evidence that this aster is revived or remains in the egg artificially incited to develop, but, as stated above, the asters appear *de novo*. Moreover, the division figure is not formed from one of them by division, but from two or more that arise separately. It appears that the chromosomes, through their attachment fibers, find these spindles and utilize them.

Concerning the nucleus, it appears that what takes place depends both on the kind of egg and on the treatment. In the ripe sea urchin's egg, where the polar body formation has been completed, the situation is relatively simple. The haploid nucleus divides; each resulting cell has the haploid number of chromosomes. It appears that the cells of the larvae also are haploid. Occasionally it seems likely that the chromosomes may divide, but the first division of the egg fails. Thus the egg becomes diploid before the next division (which is now the first complete one) takes place, giving a diploid embryo. It seems not improbable that the rare cases in which the embryos have been carried through to the adult form are those that are diploid.

Practically the same conclusion applies to the frog's egg. After puncture, the two polar bodies are given off and the haploid egg nucleus retreats to the center of the egg, where it comes into contact with two asters that have appeared in the egg, Fig. 92. The haploid embryos appear to be weak. They have at the start twice too much protoplasm for their chromosomes. Whether this is the primary cause of their weakness is not certain, nor is it known whether there would be a regulation in later stages between the haploid chromosomes and cell size. The only individuals that have been carried to the adult stages have been found to be diploid.

The known behavior of the chromosomes in natural parthenogenesis furnishes an opportunity for the study of genetic problems. Since only a single polar body is given off and the chromosomes split lengthwise, as in ordinary mitosis, all the offspring of a female line have exactly the same genetic composition. Owing to the genetic identity of these females, the amount of individual variation they show will be a measure of the slight differences of environmental influences. In fact, measurements that have been made show the same close resemblances that identical twins exhibit, and for the same reason, since in both cases the individuals have identical genes. Conversely, the sudden and often great changes, both structural and physiological, that take place when the parthenogenetic cycle ends and the sexual phase begins, furnish an opportunity to study to what extent certain new environmental influences will give a new end result with the same genetic background. Here again the same problem that has been discussed in other situations appears, namely, as to whether the change is due to different genes coming into action, or whether the product of all the genes remains the same and the environment causes their products to give a different end result. Again the answer is not convincing, but since the change has been shown to take place suddenly, much as stimuli act on organisms, rather than continuously throughout the development, it may seem more probable that different genes are brought into action at a critical stage.

CHAPTER XIV

REGENERATION

Comparisons are sometimes made between the development of a whole embryo from a fragment of an egg, and the restoration of a lost part of an adult animal. If the comparison has any value it should also be made between the failure of some isolated blastomeres to make a whole embryo and the inability of some adult animals to replace lost parts. In fact, such comparisons are not very enlightening, but nevertheless the regenerative processes in animals present in themselves some unusually interesting facts relating to development, and the reconstitution of a whole embryo from a piece of an egg may involve some of the same factors as the restoration of a whole animal from a piece of a protozoön or of a hydra.

There are two main types of regeneration. In one the new structure develops by a remodeling of the old materials; in the other the new structures are formed out of new materials that are derived from the old part. An excellent example of the first case is found in the protozoön, Stentor, Fig. 93A. If cut into two pieces, Figs. 93A , a, b, each piece rounds up, and in the course of a day or less, each remodels itself into a Stentor of half size, although there remains for a time some indication of the part from which the piece has come. This is best seen when the animal settles down and extends itself, Figs. 93a', b'.

Another example of the direct transformation of a piece into a new organism is shown by the multicellular animal, Hydra, Fig. 94A. If cut into several pieces, each piece contracts, closes its open ends, and after a few days gradually assumes the typical form of that animal, Figs. 94a–e. Small tentacles are pushed out near the anterior end, the piece becomes longer and narrower, a mouth forms at one end and a foot at the other.

A related animal, the hydroid, Tubularia, also regenerates out of
the old tissues, but since these are enclosed in a stiff chitinous wall,
the piece does not first contract and remodel itself into a new whole
of smaller size, but the new hydranth is from the beginning almost
as large as the original hydranth. Tubularia is a colonial form con-

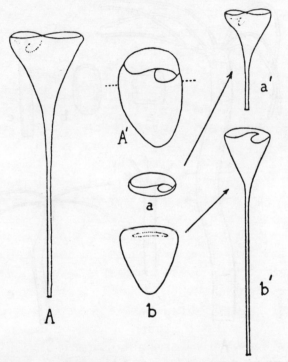

FIG. 93.   *A*, Stentor coeruleus extended; *A'*, contracted, showing level of cut into two
pieces, *a* and *b; a', b'*, regeneration of new Stentor from each piece. (After Morgan.)

sisting of a stem branching at the base from which arise tubular
stalks, each ending in a head or hydranth, having two circles of
tentacles, a peristome and a mouth, Fig. 95a. If the hydranth is cut
off, the open end of the stalk becomes covered in less than half an
hour by a plate of cells that spreads out from the cut surface of the
stalk. The covering cells are not new ones, but old ones of the wall

lying at the cut edge. In about twelve hours two circles of pigmented bands appear near the cut end, Fig. 95b, the forerunners of the two sets of tentacles. These bands pinch off from the surface except at the base to become the tentacles. Beyond the outer circle of tentacles the peristome is formed with a new mouth at its end.

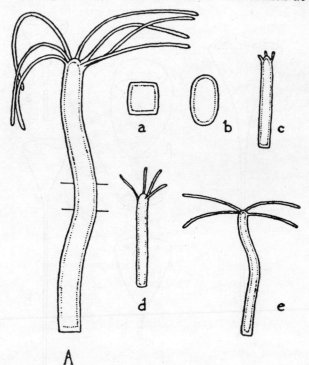

FIG. 94. *A*, Hydra viridis. A piece, *a*, cut from middle, and its development in *b*, *c*, *d*, *e*. (After Morgan.)

The new hydranth then pushes forward beyond the cut end of the chitinous tube, Fig. 95c. A histological study of the piece, when these changes are taking place, has shown that few, if any, new cells are formed; the old cells change over directly into the new parts, tentacles, peristome, etc.

The walls of the stalk of Tubularia are made up of two layers of cells inside a cuticle. There is a large central digestive tract with a longitudinal partition that divides it incompletely. The digestive fluid circulates up one side and down the other, passing around the free end of the partition just below the hydranth. Soon after the old hydranth is removed the partition breaks down near the cut surface, and its cells, bearing red pigment granules, are set free in the digestive fluid. Since the tentacle bands develop red pigment

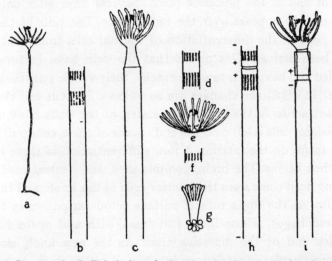

FIG. 95.   Regeneration in Tubularia. *a*, hydranth and stalk; *b*, beginning of two circles of tentacles near cut end; *c*, hydranth from last; *d, e*, regeneration of hydranth without stalk from short piece; *f, g*, regeneration of distal end of hydranth only; *h*, a short piece grafted in reverse direction on + end of long piece; *i*, regeneration of single hydranth from same. (After Peebles.)

when they appear, it was suggested that the red pigment of the digestive fluid (before its origin was known) was formative stuff; but this view overlooked the simple fact that at the time the pigmented bands are laid down the granules are circulating through the whole length of the stalk, and are not accumulating at the anterior end. It is only after the tentacles are laid down and the head formed that the granules get stuck together in the stomach of the new hydranth. They are ejected from the mouth when the hydranth pushes forward beyond the old cuticle wall. This attempt to find a formative substance was obviously premature.

As an example of regeneration by means of the development of new cells, the process in the fresh-water annelid worm, Lumbriculus, may be chosen. If the worm is cut in two, a new tail grows from the posterior cut end of the anterior piece, and a new head from the anterior end of the posterior piece. Several days after cutting, a knob of cells appears over the cut surface. The cells of the knob show none of the differentiation of the old cells from which they have been derived. This means that the cells have become more rounded and have lost, to some extent, their visible structural characters. In detail the changes are as follows: The cut end closes by the contraction of the body wall. Later, as the walls relax, a layer of ectoderm cells is left over the end. Some of these cells pull in, and begin to fill up the interior. A new differentiation of these interior cells then begins. The brain, commissures, and ventral cord of the forming head come from the interior cells of the knob, and the muscles also. At the tip, a tube of surface ectoderm pushes in to form the oesophagus, whose inner end fuses with and opens into the anterior end of the digestive tract. As the new knob elongates it shows circular constrictions into six or seven rings, or somites. The process then ends without replacing all of the anterior end (if more than seven or eight segments were removed), but enough of it to give a head.

The development of a new tail is similar, except that after the first few segments are produced there is laid down a growing zone at the tip which progressively adds new segments to those first formed, and may ultimately give rise to as many as were cut off.

The most significant features of the regeneration of the new parts in this worm concern the origin of the new cells and their fate. The new skin comes from cells of the old skin as does also the new brain and the cord and even the new muscles. As far as known these three tissues, or organs, all come from the same original cells that functioned as the skin of the worm. Their location or position in the new part would seem to determine which one of the three possibilities is realized. Where to lay the emphasis on these facts has puzzled students for a long time. At one time it was believed that like tissue produces like, but this is obviously only true in part: true in Tubularia, but true only for the skin of Lumbriculus. It has also been supposed that the regenerated organs are formed in the same way as in the embryo, i.e., from the same germ layers, but it is not true for the muscles of Lumbriculus.

This brings up once more the question of the rôle of the genes in these regenerative changes. Since all cells contain a full complement of genes it would seem that every cell is capable of forming any part of the organism, except in so far as its protoplasm has already been irreversibly changed in a given direction. On this supposition it would seem more likely that the old cells would then continue to function in the new part as before, and, in fact, this is obviously true in many cases, but in the cells that change over into different tissues this explanation will not apply. Perhaps all that can then be said is that after losing contact with their original tissues certain cells may lose their differentiation (and observation substantiates this) and then begin to develop according to their location. What "location" means here is difficult to say, but there is at least one clue. The new differentiation, in some cases, at least, starts at the point of contact with that one of the old tissues that is nearest to it. The old act as the determiners or organizers of the new, and we can perhaps assume that this is a chemical influence. It is not safe, however, to push such comparisons too far. This kind of evidence does not help to decide whether the genes are all acting while these differentiations in the cells are taking place, or whether certain sets are brought into action by the new environment of the cells in the new part.

As has been stated in Chapter I, one of the earliest attempts to explain both embryonic development and the process of regeneration is known as the Roux-Weismann hypothesis, which rests on the assumption that there is a qualitative separation of the materials of the chromosomes at each cleavage of the egg. The daughter cells come to have, in consequence, different prospective values. As already explained this was not in accord with observations known even at that time in respect to the exact, visible splitting of the chromosomes, and is still more out of harmony with the information of today concerning the genic structure of the chromosomes. Weismann attempted to bring the same hypothesis into line with the facts of regeneration. Since from any cut surface the new part is like the part removed, he assumed that, at each level of the organism, residual cells have been left as the final differentiation took place, whose "purpose" is to replace the parts distal to them. In other words: all of the cells of a given kind are not used up, but some are left in the tissues, and it is from these that the new part develops. This would also explain, in a sense, why like tissues produce their like. As a matter of fact, there is a large body of evidence that is not in accord with this view, since the cells for a new part may arise from a very different part of the body than that which gave rise in the embryo to the organs in question.

An excellent example of the last point is found in the regeneration of the lens of the eye of the salamander, Fig. 96. If the lens is removed, a new lens is regenerated from the edge of the iris. The lens is formed in the embryo from the inner layer of the overlying ectoderm. Hence the regenerated lens develops from a part of the body that has had a very different embryonic history from that of the original lens. Moreover, it develops out of the functional cells of the old iris and not from residual cells. The new lens might be said to develop from tissue that is in the best possible location to form such a lens. On the other hand, most of the cells at the edge of the iris are pigment-bearing, while the cells of the lens are transparent. In this case it may seem that if there are special genes whose activity

accounts for the maintenance of the pigmented iris cells there must
be a new set of genes that becomes more active when the lens is
removed from contact with the iris. But the result can equally well
be accounted for on the assumption that in both cases all the genes
are active, and the organ that is regenerated is primarily determined
by the new environment in which the cells of the edge of the iris

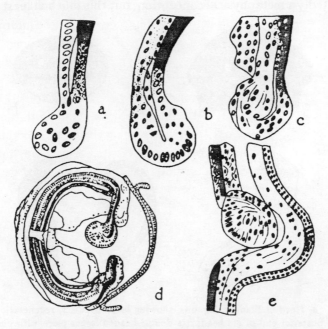

FIG. 96. Regeneration of lens of eye of Triton after removal of old lens. *a-e*, stages in
development of new lens at edge of iris; *d*, same, showing new lens in place. (After
Wolff and Fischel.)

find themselves. The protoplasm may then be considered different
from what it was before, hence with the same output from the
genes the end product is different. Of course this inference fails
to explain the specific nature of the reaction. This is the real point
at issue.

It may not be surprising, then, in the light of our inability to state what factors are at work in bringing about these results that several embryologists have resorted to ultra-naturalistic, i.e., metaphysical, attempts to explain the results. The most familiar of these interpretations passes under the name of vitalism, and in the special cases under consideration the entelechy was invoked as the vitalistic agent at work. It is perhaps futile to criticize this view so long as it is admittedly a metaphysical conception, but this much at least may be

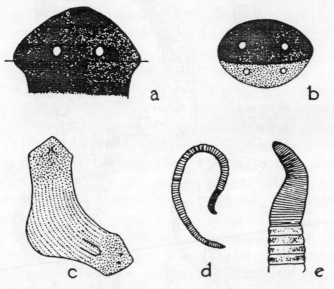

FIG. 97. *a*, Head of Planaria lugubris, showing level of cut; *b*, regeneration of new head at posterior end of old head; *c*, a double-headed larger piece with pharynx; *d*, tail of earthworm regenerating tail at anterior cut surface; *e*, same enlarged. (After Morgan.)

said: first, that the explanation is as difficult to "understand" as are the facts to be explained; second, that owing to its finalistic assumption it might, if accepted, tend to stop all further attempts to find a naturalistic explanation; and third, that there are certain facts known in regard to regeneration that put the teaching in rather a ludicrous light, in so far as the entelechy carries with it the implication of a beneficent agent.

For example: if an earthworm is cut in two in the middle, the posterior piece develops a head of four or five segments, but since the reproductive region does not regenerate, the "animal" can never reproduce itself. If the worm is cut farther back, a tail that continues to add new segments develops at the anterior end of the posterior piece, Figs. 97d, e, but the two-tailed piece dies after a while since it cannot take food.

Similar results occur when a small piece at the tip of the tail of a planarian worm is removed. It regenerates another tail at its anterior end. If the tip of the head of a planarian is cut off, a new head develops at the posterior cut surface, Figs. 97a, b, c. The two-headed piece is destined to die, since no mouth appears. It looks as though the entelechy may make grave mistakes, and since this happens consistently for a given situation, one is encouraged to continue to look for a different kind of explanation.

In the higher plants most regenerative phenomena are so different from those in animals that it is difficult to compare them. Removal of the terminal growing tip is followed by the development of some of the preformed buds that are already present lower down on the stem. Here the problem is concerned rather with the agents that hold the development of the buds in check, and there is evidence proving that this may be concerned with the development of some material formed in the tip. This material is a substance that has been isolated and its function demonstrated experimentally. It would seem, then, that the chemical substance has at least some of the properties of an entelechy.

Other plants may regenerate also by the formation of little plantlets at the cut surface. The cut surface first becomes covered with unspecialized cells from the so-called cambium layer. At one or several points of the new surface a small bud appears that probably often comes from a single cell, in other cases from several cells. One of these plantlets may grow out into a terminal stem or branch, or if removed and planted it will produce roots and become a whole plant.

In the two cases selected to show regeneration in animals by the development of new tissue, the development took place from anterior and posterior cut surfaces. Lateral regeneration also may take place, as when the leg of a salamander is cut off at any level. As much as is cut off regenerates from the stump. The cut-off piece dies for want of food. If the right leg is cut off the new one is a right leg;

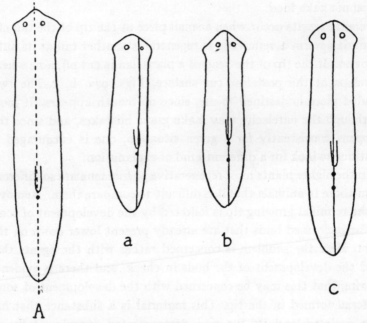

FIG. 98. *A*, Planarian cut in two lengthwise. *a, b, c*, regeneration of whole worm from left half of *A*. (After Morgan.)

and if the left leg, a new left one. If a flat worm is cut in two lengthwise, Fig. 98A, each half replaces the missing half. A piece from the left side of the worm, Fig. 98a, develops a new right side, and vice versa for a piece from the right side. A mid-plane is established· between the old and the new parts, Fig. 98c. This case is similar to what happens when the first two blastomeres of amphioxus are separated. One of them would have formed the right side of the body had the two remained together, but after isolation its materials give both right and left sides of the embryo. It does this, not by regenerating cells for the missing half, but by establishing a new mid-plane in the spherical blastula which was, so to speak, made up of right-side cells. This is what also takes place if Stentor is cut in two pieces in a longitudinal plane.

There is a problem connected with the rate of growth and the level of the cut surface that is suggestive. If the tips of the tails of two earthworms are cut off, one near the tip, the other farther forward, the rate at which the regeneration takes place is slower from the cut surface that is nearer the tip, and faster from the cut surface from farther forward. The two posterior ends may replace the missing parts in the same time, owing to the faster growth from the cut surface that is farther forward. This may be expressed teleologically by saying the greater the need the faster the growth—or that entelechy works faster at the anterior half because the need to perfect the worm is "felt" to be greater there. There is, however, a biological limit even to the "intelligence" of the entelechy, for if the cut is still farther forward no posterior growth at all takes place.

Similarly for the leg of a salamander. If the tip of one toe is cut off it will take as long to replace it as it does to replace the whole leg cut off at the base. Obviously the need for a whole leg is greater than the need for the tip of one toe, etc.

Leaving aside metaphysical interpretations, these facts concerning rate and level present some interesting comparisons. The cells in the tip of the toe of the normal, whole salamander, or at the end of the tail of the normal worm, have ceased to mutliply except in so far as to replace the normal wear and tear of the parts. It would seem, then, that whatsoever factors there are in the whole animal that restrict its possible maximum rate of growth, the same factors regulate the rate of growth of the cells in regeneration. It is not obvious, however, why farther forward the rate of regenerative growth should be greater than at the tip. Further explanation is called for.

It is well known that an annelid or a planarian will regenerate when no new food is taken in. The old, starving part actually grows smaller as the new part increases in size. But the size of the new part is markedly smaller in a starving than in a feeding animal. This relation emphasizes the fact that the amount of food available in a given region of the body is not a factor concerned in regeneration,

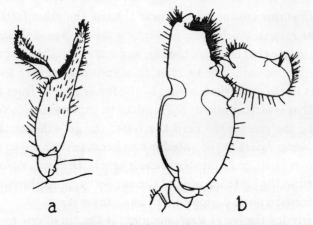

FIG. 99. *a*, Small claw of Alpheus; *b*, large claw. (After Wilson.)

FIG. 100. *A*, Young male crab at an early stage when both claws have more the form of the male claw; *B*, young crab that has lost the left claw and has regenerated a new claw on the right side like that of the female. (After Morgan.)

except in the size of the new part. The problem is more complex than this; for, in addition to the rate of growth, the kind of growth that takes place is still more significant. Its relation to the old part is clearly dependent on other things than the amount of food available.

Something of the same sort is observable in compensatory regeneration, which can be best illustrated in the case of the regeneration of the large and small claws of the crustacean Alpheus. The two claws are not only different in size, but in shape, Fig. 99. If the large claw is removed a new claw begins to regenerate at the base. When

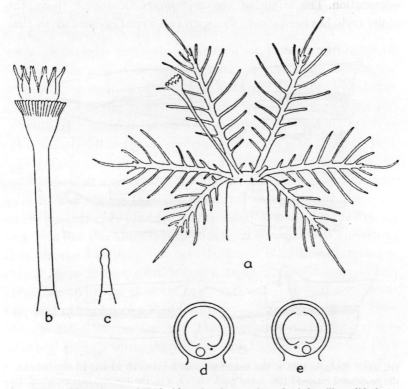

FIG. 101. *a*, Young stage of Hydroides, showing one branch of the gill modified as an operculum; *b* and *c*, functional and rudimentary opercula of adult; *d* and *e*, diagrams showing the attachment of the opercula. (After Zeleny.)

the animal molts by drawing itself out of its old cuticle, the original small claw is found to be transformed into the large claw, and the regenerated claw into the smaller one. If, on the other hand, the small claw is removed, the new claw is the small one, and the old one, after the molt, is still the large claw. It seems that here there is some sort of physiological balance between the two kinds of claws, but its nature is unknown.

In other decapod crustaceans where there is also a difference in the claws of the two sides, this transfer does not take place during regeneration. The origin of the asymmetry in one of them, the fiddler crab, has been shown to depend on a relation similar to that

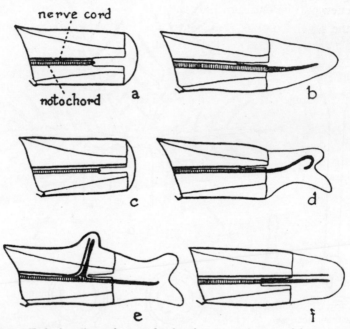

FIG. 102.   Tadpole tails. *a*, the notochord and nerve cord removed from cut end; *c*, the notochord removed (the nerve cord left); *b*, *d*, *e*, the tail regenerates when notochord reaches cut surfaces, but not when only the nerve cord reaches that surface, as in *c*. (After Morgan and Davis.)

in Alpheus, but no reversal takes place in the adult stage. The young crab has at first both claws alike, Fig. 100. If one of them is lost at the next molt it assumes the shape of the small claw, after which no reversal is possible.

In the tube-annelid, Hydroides, there is a plug, operculum, Fig. 101b, that closes the opening of the tube when the animal withdraws into it. On the opposite side there is a rudimentary plug, Fig. 101c. If the larger functional plug is removed, the smaller one becomes the larger one. The regenerated plug remains small. The reverse may again be brought about by removing the newly regenerated large plug. The conditions here are the same as in Alpheus.

Finally it has been shown that the initiation of the development of a new part is conditioned by the presence at the cut surface of one of the several organs that contribute to the part in question. If

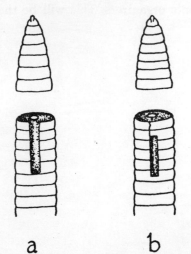

a   b

FIG. 103.   a, Showing operation of removing ventral nerve cord from cut surface; b, same in another way. New head develops only when nerve cord or piece of it (in b) at cut surface. (After Morgan.)

the notochord of the tail of the tadpole is removed from the cut surface, a new tail fails to regenerate, Fig. 102c. If, however, the notochord from its cut end farther forward extends to the cut surface, by forming new cells that later reach the exposed surface, a tail may then start to develop, Figs. 102a–f.

If the ventral nerve cord of the earthworm is removed from the cut surface a new head fails to develop at the anterior cut surface, but one may appear farther back where the old nerve cord ends,

Fig. 103. The regeneration of a salamander leg from the cut end is dependent on the presence of material derived from a bone. If instead of cutting off the foot, a slit is made in one side down to and into the bone, another secondary foot may be induced. These particular organs, nerve cord, notochord and periosteum, may be said to be organizers of the new part, since they must be present in order that a new whole may regenerate. More recently the same question has come up in connection with certain grafting experiments of embryonic organizers. This will be the topic of the next chapter.

CHAPTER XV

LOCALIZATION AND INDUCTION

The most familiar fact about embryonic development is the orderly sequence of the changes from egg to adult. This is especially obvious in the cleavage stages of the egg of annelids and molluscs where every cell can be traced to a definite organ of the embryo. So striking is this relation that it has been said that the cleavage proceeds as though directed by the end in view, i.e., purposefully. Such a purely teleological statement puts the cart before the horse from the viewpoint of science.

The theory of localization of organ-forming regions during cleavage may appear to be a modern form of the very old doctrine of preformation. The ancient doctrine postulated outright the presence of a minute invisible manikin in the egg (or in the sperm) that had only to grow big to become a man. We know too much today about what is taking place in the development of the egg to accept such a simple solution. The modern doctrine of preformation or prelocalization of organ-forming regions in the egg also tells us nothing about the way in which each region of the egg is set aside to become a definite part of the embryo, even if such a relation were strictly true. There is experimental evidence to show that this picture leaves out of account important movements of the materials of the egg, and later influences that are also as essential to the process of development as is the simpler relation of regional determination. Again, the view of prelocalized regions leaves out of account the potentialities of the egg before its median plane is determined, and it may well be that similar possibilities are present in later stages of development. To ignore this evidence may lead to oversight of some of the most essential properties of the developmental process. The formation of a whole embryo from a piece of an egg or from an isolated blastomere shows that something more than strict prelocalization of organ-forming regions is involved.

Recent discoveries have indicated that certain important phases of development are initiated by the influence of one region on another region. This information may make us pause before accepting as sufficient the statement that the problem would be solved if we could trace the organs of the embryo to definite regions of the fertilized egg. The results of the experimental work, now to be reviewed, have a direct bearing on these questions.

There are two kinds of experiments, closely interrelated, that furnish significant data. First, the artificial transfer of cells from one region of an embryo to another region of the same or of another embryo, in order to discover whether they self-differentiate there, or whether they take part in the development of the region to which they have been transferred. Second, the transfer of certain materials to other regions where they bring about the development of organs that would not otherwise appear in those regions. Most of these experiments have been made with the relatively large eggs of salamanders; others have been carried out with the small eggs of sea urchins.

When a small piece of ectoderm of triton, consisting of surface cells, is taken at an early gastrula stage from the region in front of the blastopore—a region that would normally become a part of the neural plate—and transplanted to another gastrula in a region where ectoderm would normally develop, Figs. 104d–f, it becomes ectoderm. Reciprocally, when the piece from the ectodermal region is transplanted, Fig. 104a–c, to another gastrula in the region of the prospective neural plate, it takes part in the formation of the neural plate. It is evident that *at this time* the location of the transplanted piece determines its fate, which may mean that the ectoderm is at this time not differentiated, or if so, not changed so far that it cannot be changed over by other influences when in a new position. On the

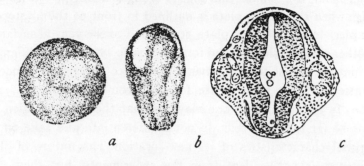

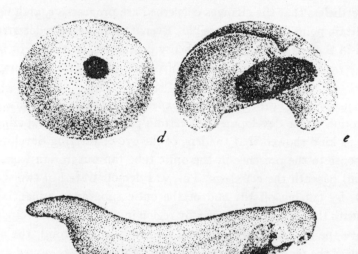

FIG. 104.   *a*, An early gastrula stage of Triton tæniatus in which a small piece of prospective ectoderm of T. cristatus has been inserted in the region of the prospective neural plate of this embryo; *b*, later stage of *a*; *c*, a cross section through the embryo that developed from *a* and *b*. The right side of the neural tube and the right eye vesicle have come from the implanted piece. *d*, An early gastrula of T. cristatus in which a piece of prospective neural plate material of T. tæniatus has been inserted in the prospective ectoderm of this embryo. In later stages, *e* and *f*, the implant has become part of the surface ectoderm. (After Spemann.)

other hand, if a similar exchange is brought about at a little later stage when the neural plate is outlined in front of the blastopore, the piece from the neural plate, transferred to the ventral surface of another embryo, sinks in and forms a vesicle whose walls differentiate into a neural tube. Conversely, a piece of ventral ectoderm, transferred to the neural plate, fails to become a part of the neural tube. In the latter case one may say that the differentiation has gone so far that the cells do not redifferentiate and take on the regional characteristics of its new location. The nature of these influences is not evident from the experiments, but they show nevertheless that the changes concerned are progressive, and, up to a certain point, may be reversible. Even if there were predetermination in the egg, as the mosaic theory postulates, it is evident from these experiments that the effect of position is at first more important.

Equally significant are other experiments that determine the direction of the development of certain parts. The first experiment of this kind showed that the lens of the eye of the frog develops in response to the presence of the optic tube (an outgrowth from the brain) beneath the ectoderm. This was demonstrated in two ways: first, by cutting off the end of the optic tube and implanting it beneath the ectoderm of a near-by region in the head. The ectoderm above the implant thickened and formed a lens; second, the ectoderm of the region where the lens would appear was removed and a new piece of ectoderm from another part of the body was placed over the exposed region. From it a lens developed. Evidently the lens develops in response to the presence beneath it of the optic tube. The lens may be said to be induced in the ectoderm by the presence of the optic tube: the latter is an inductor—or, in recent parlance, an organizer or organizator for the lens.

Evidence of the same sort has been obtained in connection with the development of the central nervous system of triton. In order to understand the bearing of the evidence, the prelocalization of the parts of the embryo of triton may be recalled. This is illustrated in Figs. 105A–E. In A, a side view of the embryo is shown, and in B, the ventral side at the time when the dorsal lip has just appeared. The material that will form the neural plate occupies the dor-

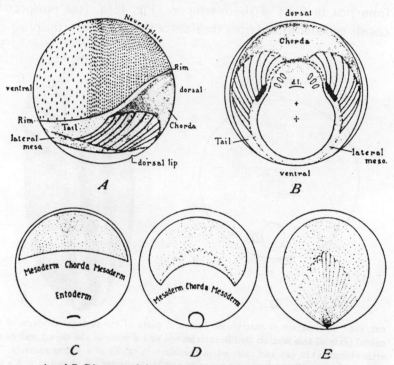

FIG. 105. A and B, Diagrams of the areas at the surface of an embryo of Triton at the time when the dorsal lip of the gastrula has just appeared; A, in side view; B, as seen from the yolk pole. (After Vogt.) The three lower figures, C, D, E, show the region of the prospective neural plate (stippled), and of the chorda and mesoderm in front of the dorsal lip. These figures also indicate the elongation of the materials of the neural plate during the closure of the blastopore when the chorda and mesoderm are turned into the blastopore lip. (After Goerttler.)

sal upper quadrant; the materials that will become the chorda endoderm and mesoderm lie in the quadrant below the neural plate. The latter materials will be turned into the interior by passing around and into the dorsal lip of the blastopore. The fate of these cell layers was discovered by marking different spots of the surface with dyes, and, in this way, following the movements as the subsequent development took place.

The most significant experiments were those in which material from just in front of the blastopore, Fig. 106a, (the prospective chorda endoderm) and from the sides of the blastopore (the prospec-

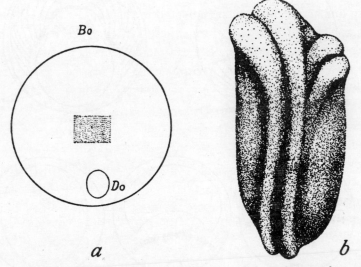

a   *b*

FIG. 106. *a*, Diagram of gastrula of Triton. A piece of the surface ectoderm of the neural plate of this neurula had been removed and a piece of the dorsal wall of the archenteron taken out and inserted in the blastula cavity of a young gastrula. The embryo that developed from this gastrula and its transplant is shown in *b*. A small secondary neural tube is present at the side of the anterior end of the primary neural tube. (After Geinitz.)

tive mesoderm) were removed and implanted on the ventral or lateral surface of another younger embryo, i.e., one still in the late blastula stage, or one about to begin to gastrulate. The implant sinks below the surface, and later above and in front of it a neural fold develops, Fig. 106b, in the ectoderm of the host. Even eyes, ears and other related structures may appear. The implant itself gives rise to notochord and mesoblastic somites. The prospective chorda mesoderm acts as an inductor calling forth in the ectoderm the development of a neural tube, which in turn may induce ears, nasal pits and other structures to appear in indifferent ectoderm; organs which would in normal development not appear in those regions. Whether the prospective endoderm or the notochord or the mesoderm cells of the gastrula stage are equally potent is not entirely clear from this experiment alone, since there is no such sharp line of demarcation as indicated in the figures between these cells at the gastrula stage. From these earlier experiments it was supposed that the induction comes from the chorda endoderm cells that normally are turned in to become the roof of the archenteron, above which the neural plate develops in the normal embryo. The same kind of reaction would then be supposed to take place in the experiment when chorda endoderm comes to lie anywhere beneath ectoderm. This simple and apparently logical inference has, however, turned out to be too limited in scope to cover the situation, for other experiments have shown that the induction is not the property of the chorda endoderm alone, but is also possessed by other cells of the early embryo. These experiments may now be considered.

The following series of experiments were made in a different way, namely, by inserting pieces from the prospective neural plate (taken at different stages) into the blastocoel of other embryos in the early gastrula stage, Figs. 107 and 108. A group of cells in the region of the prospective neural plate was removed from the roof of an early gastrula stage (at the time when the blastopore is U-shaped), and placed inside the segmentation cavity of a beginning gastrula stage, Figs. 107a, b. As gastrulation proceeds the introduced piece comes to lie against the roof or side of the recipient as the mass of yolk cells pushes in to fill up the original segmentation cavity. The piece then lies between the ectoderm on the outside, and the yolk or mesoderm on the inside. In general it may be said that the ectodermal transplant fuses with the superficial ectoderm of the host and does not act as an organizer, but in a few cases it induces a neural tube. It follows that at this stage these ectodermal cells, prospective neural plate cells, do not always bring about changes in the host, as do similar cells from a later stage.

In another series of experiments a piece of the ectoderm was removed from the prospective anterior end of a late gastrula with

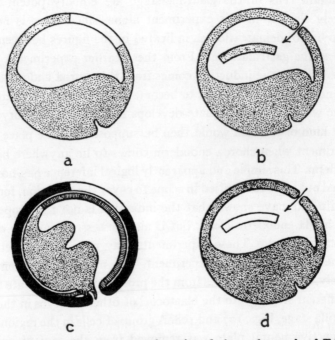

FIG. 107. Diagrams illustrating transplantation of pieces of gastrula of Triton. *a*, piece of upper wall transplanted into *b*; *c*, piece of outer layer of upper surface transplanted into *d*. (After Mangold.)

yolk plug, Fig. 107c, and transferred to the segmentation cavity of a younger gastrula, Fig. 107d. The results were variable, sometimes only a swelling appeared above the graft, but in other cases there was clear evidence of induction, as shown by the development of neural plate structures (8 out of 29 cases).

In a third series of experiments a piece was taken from the anterior corner of the neural plate, Fig. 108a, and introduced into the segmentation cavity of a younger gastrula, Fig. 108b; a neural plate develops above it in the "ventral" ectoderm of the host. In addition secondary optic cups may develop at the anterior end of the nerve tube, and a lens from the overlying ectoderm; ear vesicles may also appear; and nasal pits sometimes develop at the anterior end of the secondary embryo.

In the foregoing experiment the transplanted piece came from the anterior neural plate. The principal organs that appeared were those characteristic of the anterior end. If, however, the piece is

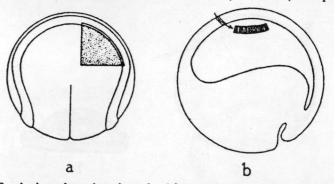

a   b

FIG. 108. A piece of ectoderm from the right anterior end of the neural plate of a Triton embryo, *a*, transplanted into the segmentation cavity of a young gastrula, *b*. (After Mangold.)

taken from the posterior end of the neural plate of the older embryo, these anterior organs are absent, and a tail region develops from the host.

There is further evidence that the response of the host depends to some extent on the position of the underlying piece, whether nearer the anterior region or farther back or at the sides. The stage of the host on which the piece is grafted is also a factor in the result. In other words: all parts of the prospective ectoderm of the host do not respond equally to the same inductors, and conversely the action of the inductors themselves depends on the stage of the embryo or the region from which they come.

The most recent experiments on amphibian organizers go far toward removing from the term organizer a vitalistic implication that may have been suggested by the choice of that word. The experiments show that the induction may be directly due to a chemical or a physical reaction between the graft and the host, for these experiments have shown that a dead transplant may bring about the induction, i.e., act as "organizator" or inductor.

When pieces of prospective ectoderm or neural plate, and even pieces of unfertilized eggs, are killed by heat (60°) and introduced into the blastocoel of a young gastrula, they induce the development

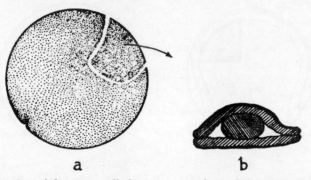

a   b

FIG. 109. A piece of the upper wall of a young gastrula, *a*, placed between two sheets of living ectoderm, *b*. (After Holtfreter.)

of a neural plate in the prospective ectoderm of the gastrula. By secondary induction eyes, ears, balancers, etc., may also develop. The results of the induction are essentially the same as when living inductors are used.

In another way the same kind of result, although not so extensive, may be brought about. When a piece from a young neural plate, or from the dorsal lip of the blastopore is killed by heat in a salt solution, and then surrounded by two sheets of prospective ectoderm, as shown in Fig. 109, a neural tube develops in the latter and later a characteristic neural structure. The presence of the inductor is essential to the process; for, if the ectodermal sheets are simply placed by themselves in the salt solution they produce only ectoderm. In another way induction may also be brought about. Pieces of neural plate or presumptive ectoderm, Fig. 110a, or even endoderm cells are killed by heat in a salt solution. On them pieces of ectoderm are laid, Fig. 110b, which develop neural tubes, Figs. 111a–b, if they remain in contact with the substrate longer than one day. These

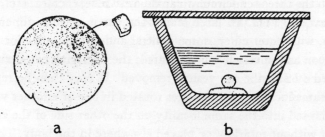

a b

FIG. 110. A piece of surface layer of young gastrula from *a*, laid on a dead layer of material from another egg, *b*. (After Holtfreter.)

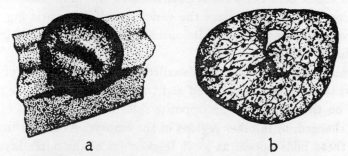

a b

FIG. 111. *a*, Formation of nerve tube in piece lying on killed material as in Fig. 110; *b*, section of same. (After Holtfreter.)

results show that the capacity to induce a neural tube is present in dead material. The simplest explanation is that the induction is brought about by chemical reaction.

THE INFLUENCE OF THE HOST ON THE GRAFT

The converse reaction, namely, that of the host on the graft, is brought out best in other experiments, especially those on the transfer of the young ear vesicles, and of the young leg buds to other regions of the embryo.

The earlier experiments are those dealing with the ear vesicles. These appear, one on each side of the head, at first as a plate or cup of cells of the inner layer of the ectoderm, that rounds up into a hollow vesicle as it sinks into the connective tissue below. From it develop the three semicircular canals, vestibule, cochlea, etc., of the inner ear, Fig. 112. In later stages the ear is a three-dimensional system, with outer-inner, dorso-ventral and antero-posterior axes.

As soon as the vesicle has appeared, the overlying ectoderm may be lifted off and the ear vesicle removed. It may then be replaced in the same location with its axes rotated in one or another way, or transplanted into the same locality on the other side of the embryo with or without rotation, or placed elsewhere in the body. The outcome is obscured to some extent, owing to the subsequent rotation of the vesicle before or while it differentiates, but in the cases where rotation has not taken place the vesicle is self-differentia ing, i.e., it develops along the lines of its own axes and is not affected by its relation to the host.

Similar problems have been studied with the limb buds of amphibians which can be removed and transplanted in the same position, on the same or on an opposite side, with one or two of the axes changed, or in other regions of the embryo. As shown in Fig. 113, these buds appear as local thickenings of the outer layer of mesoderm which forms the core of the limb. The overlying ectoderm may be lifted off and the bud taken out. It may then be placed under the ectoderm in another region, or on the other side of another embryo in place of the removed bud of that side. The possible orientations are shown in Fig. 113 (below), for the fore and the hind limbs. The circles represent the bud, and the letters inside the circle its axes. The letters outside the circles represent the axes of the host. Omitting details it may be said that the limb bud is "polarized," at the time these experiments are made, in one direction only, i.e., antero-posteriorly. This means that in whatever new position the bud finds itself, its development is in accordance with this established axis, while in respect to the two other axes its development is

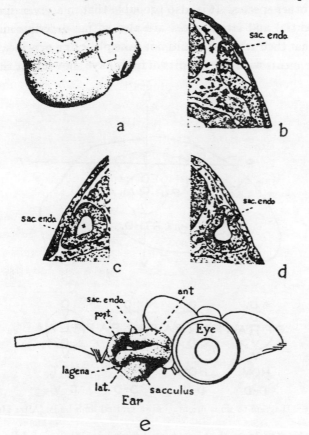

FIG. 112.   a, Young embryo of frog showing where flap of ectoderm is lifted up to expose ear vesicle beneath; b, section through forming ear vesicle; c, a transplanted ear vesicle reversed dorso-ventrally; d, normal right vesicle; e, internal ear completely formed, normal. (After Spemann and Streeter.)

determined by its new position, i.e., by influences that come from the host. Whether at the time of transposition these latter two polarizations are also present and are changed by subsequent influences from the host, or whether they have not yet developed in the bud, is not shown by the experiment. It appears, however, that in some species these axial relations may be fixed at an earlier stage than in other species. It is also probable that in a given species the dorso-ventral and lateral axes are also so far determined in later stages that their reversal would not take place in a new location.

Experiments with the gill slits of frogs have shown that the infold-

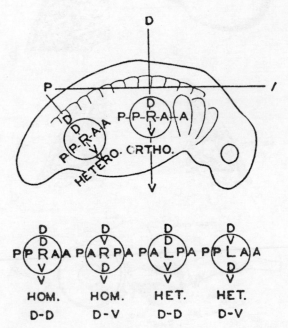

FIG. 113. Diagram to show orientation of grafted limb buds. (After Harrison.)

ings of the e toderm that meet and fuse with the outfoldings of the endoderm of the pharynx are induced by the endoderm folds. For example, when a square piece of ectoderm of a young embryo, Fig. 114, that lies in the region of the prospective gill region is removed, turn d through 180° and replanted over a similar region of another embryo from which a square piece of ectoderm of the same size has been removed, gill slits later appear in the graft over the endodermal outfoldings. This experiment is not entirely convincing because the piece is turned through 180°, which brings the vertical infoldings, although upside down, in approximately the same position as before. Moreover, in the experiment there are many indications that some induction had already taken place when the operation was performed.

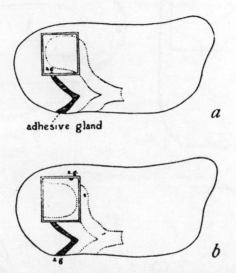

adhesive gland

a

b

FIG. 114.   Diagram showing in *a*, location of piece removed. It is turned through 180° and replaced as in *b*. The adhesive gland gives the orientation. The dotted line in both figures shows outline of future operculum. (After Ekman.)

More satisfactory results were obtained by removing a piece of ectoderm from the ventral surface over the heart, as shown in Fig. 115a, and grafting it over the exposed gill region of another embryo, Fig. 115b. Gill slits appear in the ectoderm of the graft over the underlying pouches of endoderm. Gills appear later on the arches, Fig. 115d. In addition a few small gills appear at the anterior margin of the graft, Fig. 115c, in the transplanted piece which may have been induced before grafting, since the lateral edges of the ventral piece of ectoderm may overlap the gill regions. But even here some induction from the host may be involved, since gills do not appear at the posterior edge of the same piece that lies behind the gill region.

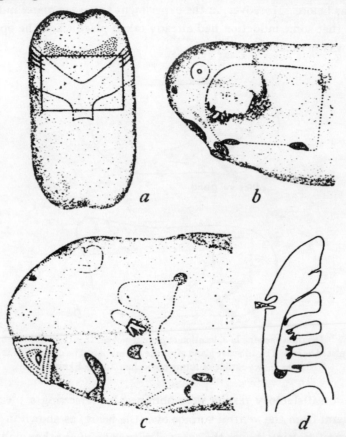

FIG. 115. A ventral piece of ectoderm in *a* is transplanted over an exposed branchial region of *b*. Gill slits are induced in transplant as seen in *d*; the gill region in *c* and *d* is covered by the operculum. (After Ekman.)

Still more convincing are experiments in which a rectangular piece, extending behind the gill region into the region of the head-kidney, is removed, turned through 180° and implanted over the same exposed region of the embryo as shown in Fig. 116. Here the region of the graft from which the gill slits would have formed now lies behind the gill region as indicated by the vertical broken lines in Fig. 116a, while the ectoderm of the graft that lies over the future endodermal pouches lay originally behind that region. Ectodermal infoldings appear later over the endodermal pouches of the gill region, and gills appear on these as shown in the horizontal section in Fig. 116c. There is clear evidence of induction in this case.

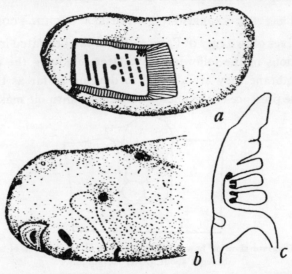

FIG. 116. A piece of surface ectoderm over the branchial region extending to level of pronephros is removed and reversed antero-posteriorly in *a*; gill slits are induced at the present anterior end of graft; the broken lines indicate the place at which the gill slits would have developed. In *b*, the operculum has covered the gill region; in *c*, the induced gill slits are shown in section. (After Ekman.)

Whether ectoderm from any part of the embryo will respond equally well in such experiments is doubtful. In one case where ectoderm from the posterior region was placed over the gill region no slits or gills developed. Here, as in some other experiments, there are indications that there is a fore-and-aft gradient, a dorso-ventral, and perhaps a radial gradient that extends outward from the prospective center for each organ, decreasing as the distance becomes greater. This may possibly be interpreted to mean that from a very early stage changes are taking place in the direction of the future development of each region. If they have not progressed too far they may be lost or reversed if a part is grafted into a new set of surroundings. The nature of these prospective influences is entirely unknown at present, and may be very different in kind and degree in different parts of the embryo.

GRAFTING EXPERIMENTS WITH SEA URCHIN EGGS

The halves of two eggs of Paracentrotus can be united in different combinations in the following way. After removing the eggs from their membrane they are allowed to develop as far as the 16-cell stage. The presence of the micromeres at the antipole makes it then

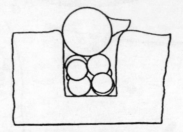

FIG. 117.   Diagram showing how grafting together of blastomeres of sea urchin is brought about. (After Hörstadius.)

possible to determine the plane in which the blastomere groups are separated. To do this the eggs are placed in calcium-free sea water ten minutes before they are to be divided with a glass needle. A photographic film is laid on the bottom of a dish of sea water, and a depression scratched in it. After separating the cells, one of the halves is transferred to the depression and placed in the desired position. The other half of another egg is then laid over the first one and a glass ball laid on it to keep the two halves in contact, Fig. 117. After a few hours they will have united and can be set free.

When halves are brought together as in Fig. 118a, the polarity being the same in both, a normal pluteus develops. Since the separation may have been either through the prospective median plane, if such be determined at this time, or at right angles to this median plane, it appears that at this time such a plane is not present, or if so that it can be changed and a new one established. When halves are united as in Fig. 118b, with polarity reversed, the anti-

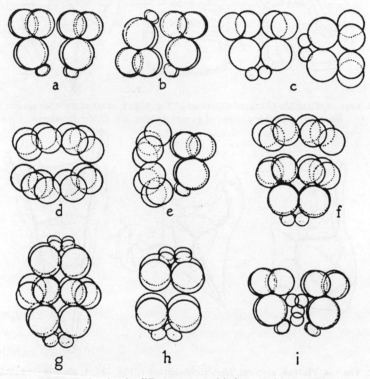

FIG. 118.   Diagram of union in different ways of halves of sixteen-cell stages of sea urchin. (After Hörstadius.)

polar (vegetative) region of each invaginates as shown in Figs. 119a and b. If as in a, a pluteus is formed with two completely separated digestive tracts, with two opposite anal openings, two mid and fore-guts, and with one common mouth opening, Fig. 119c. The skeleton is nearly doubled.

When halves are united as in Fig. 118c, oriented at right angles to each other, the gastrulation begins at two points, the two archentera unite sooner or later to form either a forked or a single gut. Either a typical or more often an abnormal pluteus develops.

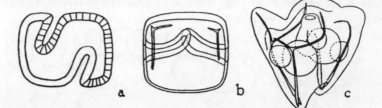

FIG. 119.   *a*, Gastrula of united halves as in Fig. 118; *b*, same of another similar combination; *c*, double pluteus of same combination. (After Hörstadius.)

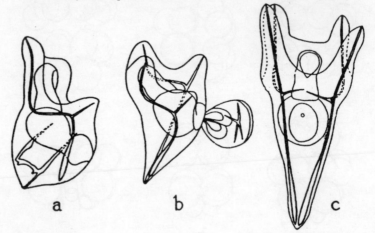

FIG. 120.   *a*, Pluteus, atypical, from combination *c*, Fig. 121; *b*, pluteus, combination *d*, Fig. 121; *c*, pluteus from combination *e*, Fig. 121. (After Hörstadius.)

When polar halves are united as in Fig. 118d, a blastula develops with two bands of cilia in opposite directions. When halves are united as in Fig. 118e (a meridional and a polar half), the combination consists of 12 mesomeres (ectoderm) 2 macromeres and 2 micromeres; a typical pluteus develops, Fig. 120c.

When a polar half is added to the polar end of a whole 16-cell stage as in Fig. 118f, the combination is two-thirds ectoderm and one-third endoderm. A normal pluteus may develop larger than the normal.

When an antipolar half was added to the polar end of a whole 16-cell stage, as shown in Fig. 118g, one pluteus was obtained with two guts and one mouth, similar to Fig. 119c.

When two antipolar halves were united as in Fig. 118h, two invaginations occurred; the larvae were quite abnormal.

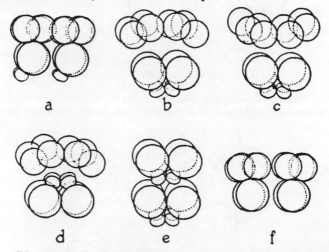

FIG. 121.   Diagram of union in different ways of halves of sixteen-cell stages of sea urchins. (After Hörstadius.)

Four micromeres were implanted in another 16-cell stage as in Fig. 118i, made up of two halves as in Fig. 118a. The number of mesenchyme cells was doubled in the gastrula stage. Typical larvae developed without doubling of the skeletal elements.

In the preceding cases the inner surfaces of the united halves are toward each other. In another set of experiments the outer surface of one half is turned toward the inner surface of the other half as shown in Fig. 121a. In order to understand how this arrangement rectifies itself, the following diagrams will be useful. In Fig. 122a, a normal blastula is drawn with the antipolar (vegetative)

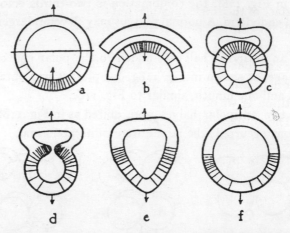

FIG. 122. Diagrams showing the results of union of vegetative (antipolar) halves inverted and united as in *b*, with animal (polar) half. (After Hörstadius.)

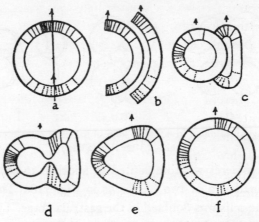

FIG. 123. Diagram showing the results of union of meridional halves of sixteen-cell stages as in *b*. (After Hörstadius.)

region striated. If this half should be cut off, turned around and its outer surface placed against the inner surface of the upper half as in Fig. 122b, both halves bend in, as shown in c. Later the two cavities unite by the withdrawal of the partition, as in d, and a single cavity results, e–f. The outer surfaces of both halves are now on the outer surface of the single blastula, but there are two centers of anti-polar cells opposite each other at the equator.

In Fig. 123a, the two halves are represented as separated along the polar axis, one is turned over and its outer surface placed inside the other half as in Fig. 123b. Both halves round up as shown in c. The partition is withdrawn, as in d, and a single hollow blastula results. The antipolar regions are on the same side of the blastula, but somewhat apart d–f.

With the aid of these diagrams the behavior of united halves of the 16-cell stage is readily understood. If two halves are united, as in Fig. 121a, both typical and atypical plutei result. If two halves are united as in Fig. 121c, both typical and atypical plutei develop. If the reverse combination is made, as in Fig. 121d, the development is quite irregular, Fig. 120b. A certain amount of regulation takes place, but the gastrulation is interfered with, and the plutei are abnormal with respect to the endoderm, Fig. 120b.

When the eight cells of two antipolar hemispheres were united with a similar orientation as in Fig. 121e, two invaginations occurred but the resulting plutei were very abnormal.

When the micromeres are removed from each of two halves which are then united as in Fig. 121f, there is no primary mesenchyme formed, but a delayed secondary mesenchyme appears. A normal pluteus may develop.

These grafting experiments show that the upper and the lower halves are already specific with respect to their future development. This was to be anticipated from the behavior of the polar and anti-polar halves when isolated, as well as from polar and antipolar halves of unsegmented eggs. There is some readjustment of the cells to each other, but on the whole the tendency to doubling, when two groups of macromeres (that give rise to the archenteron) are present, is evident. To what extent the macromeres act as inductors for the rest of the embryo is not entirely evident. The evidence points rather in the opposite direction.

CHAPTER XVI

THE DETERMINATION OF SEX

In sex determination, as in so many biological phenomena, the same end result may often appear to be brought about in more than one way. There are certain groups of animals in which there is a chromosomal mechanism which leads to the production of males and females in equal numbers. In other groups, external and internal factors other than differences in the chromosome make-up, determine whether ovaries or testes develop. In both the localization of the germ cells in certain regions of the body, whether ovary or testes, seems to be a strictly embryological problem, resting on the same basis as the localization of other organs. The question then arises whether or not it is the localization of the organ, testis or ovary, that is the initial factor determining sex, or whether it is the development of sperm cells or egg cells that constitutes the fundamental problem.

In hermaphroditic forms having both testes and ovary in different locations in the same individual, it seems apparent that there is no special problem involved beyond the same problem for all embryological localizations; and this may seem to be true also for those animals with separate sexes where the two organs are somewhat differently situated in the body. The problem is more difficult to analyze when the testis and ovary occupy practically the same position in the male and in the female. The discussion that follows considers sex determination mainly from the point of view of localization, rather than from the more usual one of the sex mechanism itself which may, however, be first briefly described.

THE CHROMOSOME MECHANISM

In several groups of vertebrates, in most groups of insects, and also in some nematode worms, sea urchins etc., there is known to be a chromosomal mechanism regulating the number of males and females. This mechanism concerns particular chromosomes known as the sex chromosomes. The female has two sex chromosomes (XX) and the male one X, and generally, in addition another chromosome that acts as the mate of X and is called the Y chromosome. This chromosome differs from all other chromosomes in that it lacks entirely, or in part, the factors concerned with sex determination. It is a dummy, and is sometimes said to be empty or indifferent, or more accurately, not to contain the kind of genes present in the other chromosomes, or at least fewer of them. The XY male produces

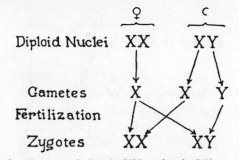

FIG. 124.   Formation of gametes in female, XX, and male, XY, types. One kind of egg and two kinds of spermatozoa.

two kinds of spermatozoa, X-bearing and Y-bearing. An X-egg fertilized by an X-sperm gives a female (XX); an X-egg fertilized by a Y-sperm gives a male (XY). The process is self-perpetuating, Fig. 124.

There is also another sex-determining mechanism that is the converse of the XX-XY one, namely, one in which there are two kinds of ripe eggs, and only one kind of sperm, Fig. 125. The female is WZ and the male ZZ. Here Z and W stand for sex chromosomes. The Z-egg fertilized by a Z-sperm gives a male; the W-egg by a Z-sperm gives a female. The W-chromosome is here the dummy. This mechanism is found in moths and birds, and in some fish.

The relationship involved in all these cases concerns the genes in the chromosomes, not only those in the sex chromosomes, but also those in the other chromosomes which are spoken of collectively as the autosomes. The evidence indicates that many genes are involved, both in the sex chromosomes and in the autosomes, and the balance between them is the determining factor for sex. Thus one may say that in the XX-XY type the X-chromosomes carry more female-producing genes, and the autosomes more male-producing genes. The Y-chromosome is indifferent. This theory is not simply an assumption based on the presence of the two X's in the female and one in the male, but is based on other evidence in which the normal

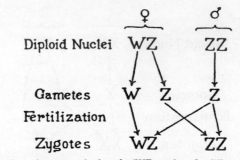

FIG. 125. Formation of gametes in female, WZ, and male, ZZ, types. Two kinds of eggs, and one kind of spermatozoön.

relations are disturbed by the occasional presence of unusual chromosome complexes, as found at times in triploids, intersexes, supersexes etc.

In the WZ-ZZ type the relation between the sex chromosomes and the autosomes is supposed to be the same in principle as in the XX-XY type, but the sex chromosomes (ZZ) are said to carry more of the male-producing genes, and the autosomes more of the female-producing genes.

The theory of the balance between the genes concerned with sex determination is the same in principle as that applied to heredity in general, where each character is regarded as the product of many genes that are present in all or in many of the chromosomes. A change in any one of the genes concerned gives a new balance that leads to a different end product. For instance, a change in a single gene gives, let us say, a new character or characters. This does not mean that the changed (i.e., the mutated) gene itself produces the character, but that it acts only as a differential in the processes that produce certain characters, because a new balance is established. Emphasis is here laid on a differential effect of one gene, or a pair of genes. It might appear probable that the difference between a male and a female is also bue to a single gene differential, but as stated above the evidence points otherwise. This statement does not mean that all the genes concerned with sex or with any other characters are equally effective. Even one of them in certain cases might turn the scale, but it appears that in sex determination the balance is the net result of the product of many genes. As a matter of fact, sex is not concerned with a single organ difference, but with a complex of differences, many of which are interrelated and interdependent.

In all cases where special mechanisms are present, the sex of the individual may be said to be determined at the moment of fertilization, although the actual development of sexual differences may not appear until a much later stage. This statement does not carry with it the implication that, even in these cases, sex may not be changed or even reversed in one or another particular, but means only that, under the ordinary conditions of life of each species the mechanism is such as to give equal numbers of males and females. Under special external and internal conditions, the developmental processes may be disturbed and the outcome be changed. Interesting as these exceptional cases may be, their occurrence does not affect the interpretation as to what happens in an environment that is normal for species, having two sexes, in which a chromosomal mechanism exists.

It is sometimes said that the female is a potential male and the male a potential female. This conveys no more information than

the facts themselves, since all individuals having all the kinds of
genes are alike, except for genic balance. Such statements are sometimes made with an air of profundity and are intended to give the
impression that "sex" is a physiological phenomenon to be studied
as such and independent of the genetic situation. If statements of
this kind are made to throw discredit on the chromosomal determination of sex, they miss the mark, for there is ample evidence
today that such a mechanism does produce males and females in
equal numbers under normal conditions. But how the genic balance
produces the results is still as much a physiological problem as is
the way in which the external environment may act physiologically
to determine the sex of the individual.

In so far as there is a difference between the genetic determination of sex and environmental determination, the latter method has
one advantage in that it may give an opportunity to attack the
problem more directly by strictly quantitative methods instead of
purely analytical ones. The recent discovery of male hormones and
female hormones in vertebrates, and the actual isolation of such
substances, may serve to bring us nearer to the genetic theory; but
the identification of these hormones with sex genes is far-fetched.
The hormones are no doubt the indirect product of genes as are all
other characters, but their effects are produced in such a round-
about manner that they may come nearer to environmental factors
than they do to the direct action of the genes on the cells in which
they are produced. Some of these substances that affect the sex
ducts or the secondary sexual characters are produced in the testes
and ovaries, if not in the germ cells themselves, but their action is
usually on other organs of the body, and in such cases may become
part of the problem of the development of those organs.

THE LOCALIZATION OF THE GERM CELLS

The germ cells, eggs and sperms, develop in strictly localized
regions of the embryo. Their origin can be traced in some cases to

the early cleavage stages, and in a few cases to a single cell, but, as has been said, the localization of the organs in which they come to lie does not appear to be different from the problem of the localization of all other organs. But whether the kind of germ cells that develop in the organs is independently determined by the balance is a further question. In a hermaphroditic animal that develops a testis in one region and an ovary in another, it may appear that the kind of germ cells is determined solely by the same factors that determine the location of the organ, but there are other hermaphrodites like the oyster in which the same gonad is alternately ovary or testis. The contrast between these situations presents problems of special interest to the embryologist, and calls for further analysis.

In a hermaphroditic animal, such as the earthworm, in which testes appear at a certain level of the animal and ovaries at another level, it may appear that the same factors that locate the testes determine that they produce sperm cells. Similarly for the ovaries. In a form like this there is presumably no chromosomal differential as in unisexual animals, but nevertheless it is legitimate to assume that a balance does exist between those genes that are sperm-producing and those that are egg-producing. This balance may be supposed to be of such a kind that it is turned one way in one organ, and the other way in another. This argument may apply also to unisexual animals where there is a sex mechanism. In some of them the location of the testis and ovary is in the same region, but whether the organ becomes an ovary or a testis may be determined by the genic balance. This last statement may be put in another way. The location of the sex organs may be the same for both, but the organ may in each case consist of two parts. The genic balance may then determine which of these parts first develops, ovary in a WZ individual and testis in a ZZ individual. Once developed, either part may produce a hormone that holds the other part in check. This interpretation can be illustrated by the development of the sex organs in birds.

In birds there is, as stated above, a sex-determining mechanism (WZ-ZZ). In the female only one ovary, that on the left side, becomes functional; that is, it produces eggs. The "ovary" on the right side remains in a rudimentary condition. If, now, the ovary of the left side is removed from a young bird, the right ovary develops into a testis, and from the base of the excised left ovary also a testis may develop. Both may produce sperm. It is practically certain that the cells giving rise to the regenerated testes in these ovariotomized birds have the same chromosome formula as had the eggs. Despite their chromosomal identity the primordial cells develop into eggs under one set of conditions in the normal development of the female, and into sperm in the same individual after ovariotomy. Furthermore, the locality is approximately the same for both. How then can we explain this apparent paradox?

There is no real difficulty in offering at least a formal explanation. If an embryo chick has the chromosome constitution of a female (WZ) at the start, the chromosome balance in the young stages determines that ovaries develop; and if the chick has the chromosome constitution of a male at the start, testes develop. Whichever organ develops first in accordance with the sex formula, its development may be said to inhibit the development of the other organ. Removal of the ovary removes the inhibition. Why, then, does not a new ovary develop rather than testes? One answer that might be given is that the balance in the new environment is in favor of the development of testes. This appears little more than a restatement of the facts, but nevertheless it directs attention to a critical point, namely, that the new reaction depends on the environment in which it produces its effects; one environment, that of the young bird, gives the ovary the advantage at the start; another environment, that of the older bird, gives the latent testis a start. This argument, somewhat more elaborated, would be as follows: At the time when the reproductive organs are first laid down in the future female, the genic balance leads to the more rapid development of that part of the primitive organ that will become an ovary. The ovary suppresses any further development of the testicular part of the primitive organ. Conversely, in the young male embryo the genic balance gives precedence to the development of the testis, which suppresses development of the ovarian part of the early gonad. When the ovary of the female is removed, its inhibiting effect on the testicular rudiment is lost, and despite the genic make up of the cells the testicular part now goes forward in its development, because, so to speak, it has already made a start in this direction.

Castrating the male bird does not affect a change in the organ that may regenerate. If any regeneration occurs new testes develop. Generally, if castration is complete not even a testis regenerates, because its rudiment is also removed. When castration is not complete a testis develops from the piece left behind. The operation on the male may be supposed to remove the original rudiment of the mass, if such is present; or else, if a rudiment of the ovary were at first present in the male it may have been completely obliterated by the development of the testis.

Theoretically the regenerated testis in the ovariotomized female should produce two kinds of mature germ cells, as do eggs with a similar chromosome composition, namely, some sperm with one X and some with a W-chromosome. It has not been shown in the bird that this is the case, but in other forms where a similar situation is brought about there is evidence that the mechanism runs true to expectation.

THE INFLUENCE OF THE EXTERNAL ENVIRONMENT ON SEX

There are several cases known where the sex of the individual is directly determined by the environment

The gephyrean worm, Bonellia, is the best known example of the influence of the environment on sex determination. The female has a body about the size of a plum, with a proboscis about a meter in length. The male is a minute wormlike form a few millimeters in length that lives as a parasite in the uterus of the female. The fertilized eggs when set free develop into free-swimming trochophore larvae, Fig. 126a. If any of these larvae settle down on the proboscis

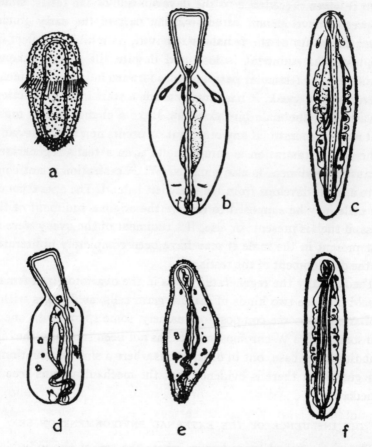

FIG. 126.　Embryos of Bonellia. *a*, indifferent larva; *b*, female larva; *c*, male larva; *d*, female intersexual larva; *e*, intermediate larva; *f*, male intersex. (After Baltzer.)

of an adult female they become males, Fig. 126c. Larvae that remain free-swimming finally sink to the bottom, and gradually develop into females, Fig. 126b. Not only is the development of ovary or testis determined by the environment, but the development of the other organs of the two kinds of individuals is also determined by external conditions in varying degrees. In many respects the male is simpler, and remains nearer to the larval form. Since it becomes sexually mature in this condition it may be classified as a neotenic form. There is no evidence that there is a chromosomal sex-determining mechanism, and if not, there is no reason for supposing that there is a sex reversal in either direction. The evidence indicates rather that in one environment the indifferent larva develops directly into a male, in the other environment into a female. Experiments have shown that it is not predetermined males that settle on the proboscis of an adult female, but that any larva will become a male in this situation. It is true that amongst hundreds of larvae kept in sea water, a few may become males—but only after some time. The result may be due to the artificial conditions that develop in the cultures: it may become more acid or alkaline, or the larvae that are starved may become males. In such cultures a few larvae also become intersexes, that is, individuals that show some of the secondary characters of the male, and some of the female, Figs. 126d–f. Intersexes may also be produced by removing at intervals to sea water larvae that have settled on the proboscis of the female.

In the normal course of events the larvae remain attached to the proboscis for about one hundred hours. They then pass into the fore-gut of the female, and at the end of two or three weeks become sexually mature males. In the first day and a half on the proboscis the anterior end of the larva shortens, loses its pigment and eye specks; on the second day the germ cells, that will become sperms, multiply rapidly; on the third day the outlet duct and its funnel appear; spermatozoa are developed by the sixth day. If the larvae are removed from the proboscis, the extent to which these changes take place depends on the length of time the larvae have remained on the proboscis. Experiments show that a sojourn of from seven to ten hours suffices to initiate the male condition. When the larvae are removed from the proboscis to water at short intervals the following results obtain: larvae that have remained only twenty to thirty hours on the proboscis, and have then been removed, become only slightly male-like in those organs that show the first response; those that have been only ten hours on the proboscis are less male-like; those remaining only four hours show later the male-like character scarcely at all. In general it may be said that after a short sojourn on the proboscis a change takes place after removal only in the anterior end, and in proportion to the length of sojourn. This relation shows that certain effects require more time than others; or, said differently, the response is different for different organs. The problem involved is not simply male versus female, but concerns each organ separately, some organs responding more promptly than others. It oversimplifies the problem to treat it as involving a change in sexuality, for many organs besides the sex cells are involved, and independently of the gonad. Under normal conditions the young animal remains long enough on the proboscis to change all of its organs into those characteristic of the male.

The experiments show that the influence of the proboscis is chemical. When pieces of the proboscis, or of the digestive tract, are dried and then soaked in sea water, the solution will change the indifferent larvae into males, or at least in the male direction. When sea water is made slightly acid ($1/2$ ccm. $n/10$ HCl + 20 ccm. sea water) the indifferent larvae are also turned in e male direction (91 6 percent). It is a fair inference, then, that chemical agents are involved, although curiously enough the secretion from the proboscis is alkaline. Furthermore, this secretion, if too strong, will actually kill the larvae. It is possible, but not particularly enlightening, to infer that both the acid, and the secretion, act as inhibitors, preventing, let us say, the development of the female organs.

There has been some speculation as to the way in which postulated male and female genes may be involved in these results. So far the suggestions have not gone beyond the purely speculative stages. The evidence from Bonellia does not seem to call for hypotheses concerning sex that are based on views concerning sex determination where males and females are the result of a chromosomal mechanism. If, as seems probable, there is no sex-chromosome mechanism in Bonellia that determines the balance of the genes after fertilization, it is entirely unnecessary to attempt to explain the results by assuming that the indifferent larvae are already either males or females.

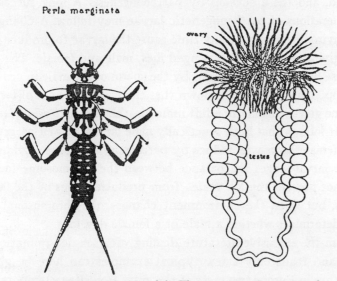

Perla marginata

ovary

testes

FIG. 127.   Larva of Perla marginata on left. The ovo-testes of a young male on right.
(After Junker.)

In other cases, where a sex mechanism has been shown to exist, there is evidence showing that it may not be effective in the early stages of development of one sex, but may in later stages lead to the development of male or female organs. For example, in one of the stone flies, Perla marginata, it has been shown that a typical ovary and testis are both present in the larval male, Fig. 127, while in the female only an ovary develops. The male is XY and the female

XX. These facts can be interpreted to mean that the chromosomal balance in the female is such as to bring about directly the development of an ovary; in the male the balance in the young stage is ineffectual, and both a testis and an ovary develop. But in the later stages of development of the male the balance, or perhaps the presence of the testis, leads to the suppression of the ovary.

Another illustration is found in one of the flies, Miastor, Fig. 128, which has the XX-XY formula. The fertilized eggs of the fly become larvae which produce eggs while still larvae; eggs only are formed, and these develop by parthenogenesis. A long succession of generations of parthenogenetic larvae may follow. A change in the environment may at any time cause the larvae to produce eggs that give rise to the adult winged flies, male and female. The eggs of these flies must be fertilized by the sperm of the male in order to develop. Now it has been shown that there are two lines of larvae: one line gives rise only to adult male flies; the other to adult female flies. It follows that both genetically male larvae as well as genetically female larvae reproduce by parthenogenetic eggs. Evidently in the larval stages the balance between the chromosome factors does not prevent the XY males from producing eggs in the larval stages, but when the environment changes, the chromosome balance determines whether a male or a female develops.

From the extensive literature dealing with sex determination in frogs and toads only a few typical examples can here be given. There are in Europe two races of the grass frog (Rana temporaria), in one of which the tadpoles develop directly into males and females. This race is called the differentiated race. In the other race all the tadpoles appear at first to be "females"—or at least they have large egg-like cells in the gonads. In this race half of the tadpoles later develop testes replacing the early egg stage, while in the other half of the tadpoles the eggs continue to develop. It is not entirely certain in this frog that there is an XY pair of chromosomes in the male, although the evidence is distinctly in favor of this condition.

If so, it may be said that the mechanism does not function as such in the early stages of the prospective male, since eggs appear first, but in later stages the XY balance directly or indirectly determines that spermatozoa are produced. Whether the transition stage of the males should be called an intersex is a matter of definition.

The embryonic development of the gonads presents some points of interest that bear on the preceding discussion. Two genital ridges appear very early on the dorsal wall of the body cavity. They contain the primitive germ cells in the surface layer. Each ridge consists of an outer cortex and an inner medulla. In later stages the eggs appear in the cortex in the females, the sperm cells in the

FIG. 128.   Miastor, sexual male, and female (to right). Three larvae with young inside (to left). (After Kahle.)

medulla in the males. Both come from the primitive germ cells. In species having the indirect development of the male, eggs first appear in the cortex. Later these disappear when sperm cells develop in the medulla. It appears then that the fate of the primitive germ cells or of their successors is determined by the part of the reproductive organs in which they come to lie.

Some interesting results that bear on these relations have come from experiments in which young tadpoles were grafted together in side-to-side union before the sex organs had developed. These parabiotic twins, as they are called, may be either a pair of males, a pair of females, or a male and a female. Since the union is made before the sex can be identified the expectation on chance combinations is one pair of males, to two pairs consisting of a male and a female, to one pair of females. When at a later stage the pairs are examined it is found that they show evidence of having been united in these proportions; but the interesting fact also appears that in some of those cases where unlike sexes have been united the development of the sexual organs may be affected. In toads neither organ is affected, but in frogs this independence is true only for the earlier stages. Later the "conflict" between the ovary and the testis becomes apparent, and the closer the twins are united the more apparent is the effect. It begins to show first in the inner members of each pair of gonads that lie nearer together than do the outer pair. The testis dominates the development of the ovary in that it tends to suppress the cortical layer of the ovary. Its suppression leads to the development of the medullary part of the ovary—that is, the part that gives rise to the sperm cells. In exceptional cases, however, the situation is reversed; and in these it is supposed that the ovary was, at the start, further ahead, or else underwent a more rapid development than the testis. Here the development of the medullary part of the male organs of the co-twin is delayed. These results give the impression that development in the two parts of the young larva are competing; the development of the cortex suppresses the development of the medullary part of the organ, and vice versa. The early development of one or the other is finally determined in the normal tadpole by the genetic composition of the individual: if it is a male (XY) the medullary layer has the advantage, except at the beginning; if a female, the cortical layer. But these normal relations may be upset by environmental conditions that may bring about partial sex reversal. This is indicated by the grafting experiments just described. Similar results can be brought about by temperature. For example: if the young tadpoles of the wood frog are kept at a temperature of 20°C., the development of the testes and ovaries is direct; the two sexes appear in a one-to-one ratio. But if the young tadpoles are transferred to a higher temperature (32°C.), the testis continues to develop normally but the ovary is affected, the cortex stops its development and its eggs degenerate. The medulla than starts to grow rapidly. The characteristic sex cords of the testes develop, and finally the sperm cells of the male appear. It follows that the suppression of the outer layer permits the development of the interior part of the sex organ, which leads to the development of testes. Here we get an insight into the relation between the development of the male and the female sexual organs. In most cases one or the other starts first, according to the genetic make-up of the individual, but this time element may be upset by environmental conditions. If the cortex develops first, an ovary results; if the medulla develops first, a testis results. The development of either suppresses the development of the other.

From another source there is experimental evidence that in the frog the female is XX and the male XY. Rarely an adult individual is found that has both ovary and testis as parts of the same gonad. These frogs are genetic females. In one case it was possible to fertilize the eggs of a differentiated race of the grass frog with the sperm from such a hermaphrodite. All the offspring were females. If the hermaphroditic female is an XX-female, that is, a genetic female, then every ripe spermatozoön should carry one X-chromosome. The eggs of the female used in this experiment should also each carry one X; hence all the offspring should be expected to be XX or females. The eggs of the hermaphrodite were also fertilized by sperm from a male of the differentiated race. Males and females were produced in equal numbers. This again is in accord with the XX-XY formula. When the eggs of the hermaphrodite were fertilized by sperm from the same individual, nearly all the offspring were females, except a few that were intersexes.

A curious fact has long been known in cattle which can now be explained on the basis of the experimental results with parabiotic

twins. When twin calves are born, one of which is a normal male and
the other a female, the latter is generally sterile. It is called a freemartin. The external genitalia of the freemartin are female—or at
least more femalelike than malelike, but the reproductive organs
resemble testes. It has been shown that each of these twins comes
from a single egg, and that later the blood systems of the two
embryos are in communication through connections established
between the embryonic envelopes, Fig. 129. The evidence shows
convincingly that the freemartin started as a genetic female, and
that the connection with the male embryo through the circulation

FIG. 129.   Two embryo calves, one of which will become a freemartin, whose placentas
are united. (After Lillie.)

has suppressed the development of the ovary, under which circumstances the testicular structures develop. Sex is partially reversed.

The problems of "intersexuality" have been extensively studied
in the gypsy moth. In different parts of the world—Europe, Asia
and Japan, there are different races of this moth. Within such races
the number of males and of females is equal. Presumably there is
a sex chromosome mechanism present as in other moths, the female
being WZ and the male ZZ. When a female of the European race is
crossed to a Japanese male, equal numbers of males and females are
produced, but if the cross is made the other way the sons are normal,
the daughters are intersexes. These intersexes are mosaics of male

and female parts. Different races of Japanese moths crossed with the European races, or with one another, produce different combinations of sexes and intersexes. In one series of crosses the females were finally all changed into males. These are female intersexes. In other series of crosses the males were finally changed over into females. These are male intersexes. The results are explained on the hypothesis that in these different races the sex balance is differently constituted. There are strong female races and weak races, etc. There is some uncertainty as to where this balance exists—whether between the sex chromosomes and autosomes, or between the W and Z chromosomes, or even between the cytoplasm and the entire chromosome group.

The mosaic character of the intersexes is explained by Goldschmidt on the hypothesis that the amount of male and of female enzymes, present at the time at which the male or female parts are laid down in the embryo, determines the results. For instance, in certain combinations of the sex factors of these racial hybrids more of the male enzyme is present at first, and therefore the individuals start as males, but in later stages more of the female enzyme is present, overtaking the production of the male enzyme, hence the later organs are femalelike. The reverse results occur in other combinations. This is, in a sense, a restatement of the facts observed, interpreted in terms of sex enzymes that develop in the cells of the part affected and not as in the vertebrates in the sex glands themselves.

This hypothesis brings up in rather acute form the question as to whether all the genes are acting all the time, or whether some of them are more active at one stage and other genes at other stages. If we assume that all the genes are active from the beginning, then the genic balance and resulting enzyme ratio in the normal female embryo (WZ) is such that all the organs laid down throughout development are female ones; and in the male embryo (ZZ) all the organs are male from the start. But in the hybrids of the gypsy moth a new genic balance is present, inducing a change in enzyme ratio and in sex characters as development proceeds. Two or three alternative explanations, then, become possible: (1) On the view that all the genes are acting all the time it would be assumed, in the normal races, that when a certain genic balance is present all organs that differ in the two sexes are affected in the male direction, and that when another balance is present the same organs are affected in the female direction. But in the hybrid combination the first organs laid down respond in one direction and later organs in the opposite direction to the same balance. Here the emphasis is laid on the kind of response of the different organs at different stages of development; a response, be it noted, to substances in the cells themselves that are affected. The genic balance is assumed to remain the same throughout, and the changing result depends on the specific responsiveness of the somatic organs and even of the sex organs to something manufactured in the cells. (2) On the assumption that some of the genes are more active at one stage of development and other genes at another stage, either in amount or quality, it would be assumed that in the hybrids one or the other set of genes becomes more active in the early stages, i.e., predominates, and at later stages genes of the other set become more active. In terms of enzymes, the amounts produced are themselves affected by the stage of development of the embryo. (3) There is a third possibility, which is a sort of combination of the two, namely, that the results depend directly on the rate at which the two enzymes themselves develop in different cases. In certain hybrid combinations the male enzymes are first produced in excess and male organs are laid down; later the female enzymes overtake the male enzymes and turn the development in the female direction. The results depend, then, on the rate or speed at which these two enzymes are produced. All the genes are at work all the time, but in the hybrids the time at which the male or the female enzymes are in excess is different in different combinations. Here emphasis is laid on the rate of enzyme development, irrespective of the relative condition of the embryonic stages. On the first view (1) the emphasis is laid on the response of the early and late organs to the amount produced by the genes; on the second view (2) the change depends on the time at which the contrasted sets of genes become more active; and on the third view (3) on the rate at which the one or the other enzyme is produced, which is determined not by different genes but by the rate of development of the enzymes in the embryo. There is at present no way of deciding amongst these possibilities. Any one of them may be imagined as a physiological process, but there are so many unknown factors that to claim any of them as a dynamic theory of sex seems rather extieme; for dynamics, in the physical usage of the term, deals with measured quantities and known physical properties of the agents appealed to.

CHAPTER XVII
PHYSIOLOGICAL EMBRYOLOGY

In the preceding chapters almost all the emphasis has been laid on the irreversible changes in form that take place during development, because these changes are its most visible and best known characteristics. Along with these changes in form occurs the elaboration of particular kinds of substances. Both the changes in form and the chemical changes may be regarded as physiological processes. The interrelation of the two brings us face to face with some fundamental considerations. Physiology, as generally understood, is concerned with the recurring functioning of organs which takes place without any permanent change in their form or structure. The question arises, whether the changes occurring in the embryo involve something more or something different from the physiological functions of adult organisms.

It is not safe, however, to take for granted that the adult animal itself is structurally in a static condition. It is known that many of the organs of the adult animal are continually renewing themselves; the skin is wearing away at the surface, and its cells are replaced by the new cells from the deeper layers of the skin; the glands seem to replace the cells that have functioned for a time; the blood cells are being continually destroyed and renewed. The most permanent cells that persist throughout life are those of the central nervous system, muscles, and bones, but there may be occasional renewal of bone, and after injury its periosteum, or outer layer, shows remarkable powers of regenerating a new bony structure. Nevertheless, the contrast remains: the adult organs do retain their form; the embryo undergoes rapid changes in form.

In other respects there are obvious resemblances. Respiration takes place in both; oxygen is taken in and carbon dioxide is liberated; water plays an important rôle in both; carbohydrates are used by both for some of the chemical changes that take place, and the same statement may be made for fats and oils, and for proteins. The adult gets these materials from outside and digests and metabolizes them, but they are already present in the egg, and the transformation of the proteins, fats, and carbohydrates probably involves the same kinds of chemical changes in eggs as in adults. Their transformation is due to the presence of enzymes which, acting as catalysts, transform the materials, supplying the embryo with the energy required for its transformations. The stored materials of the egg are also available as food for the rapid growth of the embryo, which means that they are changed over into protoplasm. Many salts are essential for development. In marine embryos, development takes place in sea water, and it has been shown that the embryo takes up some of its salts directly from the sea. In eggs that develop in water, both fresh and salt, water also is taken in. In land animals, especially insects, there is enough water in the egg to carry the embryo through its early stages. Its loss by evaporation is prevented by the membrane around the egg. Finally, the presence of vitamins in the egg is practically certain. These, while present in minute quantities, suffice to carry through the early stages. Whether the egg or embryo can manufacture them from its own materials, as do many plants, is problematical, but it is safe to assume that the embryo has enough to carry it on until it can fend for itself. In later stages the necessary vitamins are, no doubt, taken in with the food as in adult animals.

There is, then, a sufficient body of evidence showing that there is a chemistry of development comparable to the chemistry of the adult in many respects. The question still remains whether we are justified in supposing that a knowledge of this chemistry would suffice to account for the changes in form that the embryo undergoes. It would be hazardous to assert this and equally so to deny it, if for no other reason than that the physical properties of the egg and the embryo also play a rôle in development. Moreover, the egg itself has a definite if labile organization of its own which furnishes the basis for its development.

It is a noteworthy fact that visible building-up processes of development can be reduced to a few changes that repeat themselves over and over again. The division of the egg has all the features of cell division of the tissue cells of adult animals, and of many of the unicellular forms. The dissolution of the nuclear wall, the resolution of the chromatin network into its constituent chromosomes, the mitotic figure in the cytoplasm, etc., occur in the cell divisions of both. In one respect the division of the egg differs from that of tissue cells, namely, in the rapid sequence of the divisions, but even here the resemblance is close, since there is a resting nucleus between successive divisions. The bacteria alone a proach the rapidity of successive divisions of the egg if the medium supplies abundant food. There can be no doubt that the egg divides rapidly because it has in itself a store of the essential materials in a form ready for utilization. In another respect the cleavage of the egg differs from that of the body cells; there is no increase in size of the daughter cells to the original size before the next division. This, too, may be safely ascribed to the abundance of the materials that incite the divisions and supply the necessary energy. This statement applies especially to the earlier divisions of the egg, and may be less true for later stages.

The flattening of the cells of the segmenting egg against each other after each division, can be imitated by oil drops pressed together at the periphery. Surface tension plays a rôle, no doubt, in both cases, but if the wall of the compressing vessel is removed from the oil drops they float apart; if the fertilization membrane is removed from the egg the blastomeres still stay together. Something in addition to surface tension is therefore involved in the latter; something in the membrane that sticks them together and possibly attracts like cells to one another. It has been shown that in calcium-free sea water the blastomeres do not stick together, and may be shaken apart. This means that the presence of calcium in the sea water is essential for holding these cells in contact.

The amoeboid movement of individual cells that characterizes a few of the developmental processes resembles the movements of the white blood corpuscles of adult animals, and to all appearances is the same as that in free-living amoebae. Many suggestions have been made concerning the physical aspects of amoeboid movement, and while it cannot be claimed that this is fully understood, it is probable at least, that such movements involve changes in the surface tension of the cells.

Many examples of inturning or buckling of cell layers in the embryo are known. The process of invagination or gastrulation is an excellent example. The large cells of one hemisphere of the hollow blastula turn into the interior, Figs. 33a, b, c, when a certain stage is reached. If the individual cells, turning in, are examined each is found to have a broader interior end and a smaller outer one. This might happen if they were pushed in from outside by an extraneous force, but since no such agent can be appealed to, it appears that the individual cells change shape at this time. The broadening of their inner ends recalls the spreading out of the advancing end of an amoeba, suggesting that surface-tension is also here at play. This might be accounted for on the assumption that the increase of carbon dioxide or some other substance, by lowering the surface tension on the inner end of the cells, causes them to broaden out. Hence the wall turns in as a purely physical process. It is true that this implies that the cell is homogeneous throughout, which is probably not the case. There may be, and no doubt are, differences in the inner and outer ends of the cells that go back to the egg structure, but even so the stimulus that acts on the inner ends may be a chemical agent.

Many examples of invagination and evagination are known.

Pouches appear in several cases from the wall of the primitive archenteron; the optic tubes push out from the anterior end of the brain; the lens cups and the ear cups separate from the side of the head, etc. All these appear to be brought about by the same kind of change.

Local thickenings may appear in the walls of some of the first formed organs. These are caused by local increase in the number of cells, accompanied by the movements of individual cells away from the surface. The process is much the same as the inturning of layers. Here the cells act individually instead of as a layer, which seems to be only a variation of invagination, or vice versa.

The inturning of the neural plate to form a tube—the central nervous system of the vertebrate—is another example of inturning along a line instead of radially around a center as in gastrulation. The cell thickening is due, in part, to the pulling in of cells from the surface layer, and at the same time the plate as a whole rolls in or invaginates. The movements of the whole appear to be brought about in the same way as in other cases of invagination.

The discovery that the invagination of the central nervous system can be brought about, in regions other than those where it normally appears, by the presence of an organizer beneath the surface, even when the organizer consists of dead material, goes far toward showing that the process is a combination of chemical stimulus and physical reaction.

The preceding cases are fair samples of the most characteristic processes in embryonic development involving a change in form. On examination there seems to be nothing mysterious about them, i.e., nothing that may not involve physical principles, which do not appear to be beyond the reach of possible experimental analysis. The embryologist will be content, for the time being at least, when he can definitely demonstrate the sufficiency of physical explanations for such changes in form.

But this is only a part of the story of development and this picture is probably too simple, since it leaves out of account the self-regulations that are characteristic of each event. It ignores, too, the constitution or possible structure of the egg which may still be an important initiating agent that antedates these later changes. This must now be considered.

It might be possible, perhaps, to give a purely formal account as to how a very simple organization in the egg might be the basis for the more complex organization that develops out of it, but such an attempt is at present necessarily imperfect and speculative. For example: Around its polar axis the materials of the egg are visibly stratified. A plane of bilaterality may be supposed to be introduced by the entrance of the sperm at one point. Cleavage divides these materials into cells, some containing more, some less, of the different materials. These differences in the materials may be supposed to affect the subsequent development of different regions. The picture is, however, obviously insufficient to cover other facts now known concerning development, for pieces of the egg may be removed, yet whole development takes place. If one postulates a fixed organization of the egg of such a kind that it gives a whole organized embryo, then half of it should give only half. It seems that strict prelocalization is not sufficient to allow the egg to do the things it can do. The postulated organization must be more fluid or flexible, capable of readjustments; hence it has become the habit of some embryologists to speak mysteriously of a self-organizing power in the egg without attempting to state what kind of power can do such things. The problem is only restated. It does not help the matter much to refer such an imaginary organization to the molecular constitution of the egg. No one will deny that there may be definite arrangements of molecules in the egg, but until such ideas can be experimentally tested and shown to be important in the regulative processes that take place in the egg, as well as in regeneration of the adult, it does not help very much to postulate them. If the cells are taken as the units of development there is evidence indicating that the formation of many organs seems to involve whole groups of cells in coördinated movements. Should organization ignore the cell boundaries? It cannot be overlooked that there are many cases known where cell boundaries are strictly observed.

Centrifuging the egg has shown that its visible materials at least are not organ-forming, and, since the argument above is based on their visible stratification and movements, some embryologists have resorted to the assumption that the organization exists in the "ground substance" of the egg—the clear protoplasm in which we can see no differentiation. Such an attempt to evade the problem is mere subterfuge and not clarifying. It simply removes the problem to an invisible region, and adds nothing for our information. If the organization does exist in the ground substance, we can drive through it the yolk, oil, and pigment, without destroying it. We can cut it up into pieces and yet get a new whole from each part. One may say of course that the piece reorganizes itself, but how?

Despite these difficulties there is evidence that progressive changes, that are more or less irreversible, take place as development proceeds. It has been shown many times that cells and regions not yet visibly differentiated continue, when isolated, to behave as parts. What is the nature of these changes that are not reversible? The answer to this question would seem to be just as important for an understanding of normal development as the reversibility or reorganization of isolated pieces of an earlier stage just discussed.

What, it may be asked, have strictly physiological changes to do with the visible changes that we speak of as development? Is the sequence of the physical changes in form the direct outcome of chemical changes, or, to state the opposite point of view, do chemical changes only serve to maintain the life of the organism (as they do that of the adult) while it passes through the developmental stages? Put in another way the last question may mean that the chemical changes are mainly concerned with making available the materials of the egg for food and supplying the energy for the physical events of development. While such a distinction may be arbitrary and perhaps misleading nevertheless the contrast may be worth a brief examination.

The dissolution of the wall of the germinal vesicle of the egg that takes place in some eggs before fertilization, and in others after fertilization, would appear to be a chemical event, or possibly to be a physical change from gel to sol. When it occurs, after the sperm enters, the latter may release some substance in combination with the substance of the egg that directly causes the change; but when the wall dissolves before the egg leaves the ovary the event would rather appear to be due to an automatic process in the egg itself when it reaches a certain phase of its development. In this respect the disappearance of the nuclear membrane is comparable to the same change that takes place before the division of each blastomere during cleavage. The dissolution may be brought about by something set free within the nucleus, or it may be the result of changes taking place throughout the whole protoplasm.

The setting free of the nuclear sap and its mixing with the protoplasm appear to be simple physical changes, which, by altering the viscosity of the protoplasm, lead to a series of events that includes the movement of the spindle toward the pole. During this time the configuration of the materials of the egg, in the sense of the arrangement of its visible materials, is known to change. This leads to the next step in development. These movements, purely physical, appear to result from the setting free of the more fluid nuclear sap.

The puncture of the surface of the egg by the tip of the spermatozoön brings about instantaneous changes in the surface of the egg. It has been suggested that the puncture itself, by changing the tension of the membrane, leads to the next phase; but it seems more probable that a localized chemical event also takes place, since, at the point of puncture, extensive changes are seen to occur at the surface of the egg. The fertilization cone pushes out and engulfs the sperm head.

At the time of the formation of the fertilization cone a membrane lifts off the surface of the egg, the so-called fertilization membrane. Its origin has been much discussed. According to some observers it is present before fertilization and is lifted from the surface by the accumulation of fluid beneath it; according to others the membrane is not present before fertilization, but is formed at the moment of fertilization by a chemical change in the surface. It has been shown that at this time the permeability of the egg is changed and it has been suggested that it is this that starts development. On the other hand, since it has been shown that the egg of the sea urchin may be fertilized and may develop without the lifting of a membrane, it is obvious that the changes in the egg after fertilization may take place without the membrane being formed or at least without its being lifted from the surface.

These initial changes give many indications of being both chemical and physical reactions without involving up to this point any striking changes in the form of the egg as a whole. Inasmuch as the lifting of the membrane, followed by the cleavage of the egg, can be brought about in some eggs by purely chemical changes in the sea water and in other eggs by such physical means as heat or puncture, it seems not unreasonable to infer that the reaction can be explained in chemical or physical terms or both, and does not call for extraordinary ultraphysical explanation. The movements of the egg pronucleus (after the extrusion of the polar bodies) and the simultaneous movement of the sperm nucleus toward the center of the egg can be safely ascribed to a physical reaction as a result of the swelling of the two pronuclei which carries them toward a common point. This seems to be an adjustment due to the increasing size of these bodies in a fluid medium containing spheres of different sizes. In other words there does not seem to be any need here to assume mysrious attracting forces acting at a distance.

So far the changes involve movements within the egg itself. The next changes are visible changes in the form of the system as a whole.

Before and during the division of the egg into two parts extensive movements take place. These are visible at the surface of the egg and in the underlying protoplasm. They are synchronous with the formation of the mitotic spindle in the interior of the egg, which, amongst other changes, involves the transformation of a considerable part of the colloids of the egg from sol to gel. While we know that we can suppress the cleavage of the egg by chemical or by physical agents, and that within limits the change in rate of the successive divisions, in response to changes in temperature, conforms approximately to that of chemical and physical reactions, the processes involved in cell division are too complex to permit a strictly chemicophysical explanation. Nevertheless it is significant to note that there have been almost no attempts to refer these events to any other field than that of a mechanistic one.

The growth of the larva and of the foetus may be defined as the conversion of the stored food materials into protoplasm. A great part of the work in chemical embryology is based on this assumption. The physiology of development is, to this extent, essentially the same kind of physiology as that of the adult; the difference being due in large part to the storage of the necessary materials in the egg itself for its first stages of development, and to their origin from outside in the later stages as in the adult.

This brief review suffices to show that the embryologist is justified in his expectation that many of the problems of development may in time be brought within the scope of chemical and physical laws. It is true that these physical events can be traced in most cases to changes within the individual cells and that there is here much that is still unexplained. As a matter of fact most of the changes that take place at the beginning of development are within cell boundaries. The egg is a cell set to go along a known path, and here we have an opportunity such as is found nowhere else to follow at first hand the physiology of a cell as an individual. The physiology of the adult animal is concerned in the main with groups of cells and the interaction of their products with the rest of the organism, but there is no reason to suppose that what takes place in the individual cells of the organs is different in kind from what takes place in the single egg cell.

As I have already pointed out, there is an interesting problem concerning the possible interaction between the chromatin of the cells and the protoplasm during development. The visible differentiation of the embryonic cells takes place in the protoplasm. The most common genetic assumption is that the genes remain the same throughout this time. It is, however, conceivable that the genes also are building up more and more, or are changing in some way, as development proceeds in response to that part of the protoplasm in which they come to lie, and that these changes have a reciprocal influence on the protoplasm. It may be objected that this view is incompatible with the evidence that by changing the location of cells, as in grafting experiments and in regeneration, the cells may come to differentiate in another direction. But the objection is not so serious as it may appear if the basic constitution of the gene remains always the same, the postulated additions or changes in the genes being of the same order as those that take place in the protoplasm. If the latter can change its differentiation in a new environment without losing its fundamental properties, why may not the genes also? This question is clearly beyond the range of present evidence, but as a possibility it need not be rejected. The answer, for or against such an assumption, will have to wait until evidence can be obtained from experimental investigation.

REFERENCES

GENERAL

Beer, G. R. de. 1924. Growth. London.
—— 1926. An introduction to experimental embryology. Oxford.
—— 1930. Embryology and evolution. Oxford.
Brachet, A. 1917. L'Oeuf et les facteurs de l'ontogénèse. Paris.
Conklin, E. G. 1917. Heredity and environment. Princeton.
Driesch, H. 1908. The science and philosophy of the organism. London.
—— 1922. Geschichte des Vitalismus. Leipzig.
Dürken, B. 1928. Lehrbuch der Experimentalzoologie. Berlin.
Fauré-Fremiet, E. 1925. La Cinétique du développement. Paris.
Goldschmidt, R. 1927. Physiologische Theorie der Vererbung. Berlin.
Gray, J. 1931. A text book of experimental cytology. Cambridge.
Guyénot, E. 1930. La Variation et l'évolution, I, II. Paris.
—— 1931. L'Hérédité. Paris.
Hertwig, O. 1906. Handbuch der vergleichenden und experimentellen Entwicklungslehre der Wirbelthiere. Jena.
Hogben, L. 1930. The nature of living matter. London.
Hogben, L., and F. R. Winton. 1924. Comparative physiology. London.
Huxley, J. S. 1932. Problems of relative growth. London.
Jenkinson, J. W. 1909. Experimental embryology. Oxford.
Korschelt, E., and K. Heider. 1900. Embryology of invertebrates. I–IV. [Translation.] New York.
Loeb, J. 1916. The organism as a whole. New York.
MacBride, E. W. 1914. Textbook of embryology. I. Invertebrata. London.
Marshall, A. M. 1893. Vertebrate embryology. London.
Morgan, T. H. 1927. Experimental embryology. New York.
—— 1928. The theory of the gene. New Haven.
Needham, Joseph. 1931. Chemical embryology. I–III. Cambridge.
Roux, W. 1895. Entwicklungsmechanik der Organismen. Leipzig.
Russell, E. S. 1930. The interpretation of development and heredity. Oxford.
Schleip, W. 1929. Die Determination der Primitiventwicklung. Leipzig.
Sharp, L. W. 1926. An introduction to cytology. New York.
Weismann, A. 1902. The germ-plasm. [Translation.] New York.
Wilson, E. B. 1928. The cell, in development and heredity. New York.

CHAPTER I

INTRODUCTION

Driesch, H. 1894. Analytische Theorie der organischen Entwickelung. Leipzig.

———— 1901. Die organischen Regulationen. Leipzig.

———— 1905. Der Vitalismus als Geschichte und als Lehre. III. Leipzig.

Haldane, J. S. 1929. The sciences and philosophy. Glasgow.

Morgan, T. H. 1932. The scientific basis of evolution, Chap. XII. New York.

Needham, Joseph. 1931. Chemical embryology, I, Part III, Sec. 3, and Epilegomena. Cambridge.

Whitehead, A. N. 1927. Science and the modern world. Lowell Lectures, 1925. New York.

CHAPTER II

DEVELOPMENT AND GENETICS

Boveri, Th. 1902. Über mehrpolige Mitosen als Mittel zur Analyse des Zellkerns. Verh. Phys.-med. Ges. Würzburg. N. F. XXXV.

———— 1905. Über Doppelbefruchtung. Sitzb. Phys.-med. Ges. Würzburg.

Bridges, C. B. 1930. Haploid Drosophila and the theory of genic balance. Science, LXXII.

———— 1932. The genetics of sex in Drosophila. Sex and internal Secretions. Baltimore.

Goldschmidt, R. 1928. The gene. Quart. rev. biol., III.

McClung, C. E. 1924. The chromosome theory of heredity. General cytology. Chicago.

Muller, H. J. 1929. The gene as the basis of life. Proc. Intern. congr. plant sciences, I.

———— 1929. Heritable variations, their production by x-rays and their relation to evolution. Smithson. Rep., Publ. 3048.

Muller, H. J., and L. M. Mott-Smith. 1930. Evidence that natural radio-activity is inadequate to explain the frequency of "natural" mutations. Proc. Nat. acad. sci., XVI.

CHAPTER III

THE EGG AND THE SPERMATOZOÖN

Ballowitz, E. 1908. Untersuchungen über die Struktur der Spermatozoen .... Arch. mikr. Anat., XXXII.

Carothers, E. E. 1917. The segregation and recombination of homologous chromosomes as found in two genera of Acrididae. Jour. morph., XXVIII.

Chambers, R. 1933. The manner of sperm entry in various marine ova. Jour. exp. biol., X.

Just, E. E. 1922. Initiation of development in the egg of Arbacia. I. Effect of hypertonic sea water in producing membrane separation, cleavage, and top-swimming plutei. Biol. bull. XLIII.

—— 1930. The present status of the fertilizin theory of fertilization. Protoplasma, X.

Lillie, F. R. 1911. Studies of fertilization in Nereis. I. Cortical changes in the egg. II. Partial fertilization. Jour. morph., XXII.

—— 1912. Studies of fertilization in Nereis. III. The morphology of the normal fertilization of Nereis. IV. The fertilizing power of portions of the spermatozoön. Jour. exp. zoöl., XII.

—— 1913. Studies of fertilization. V. The behavior of the spermatozoa of Nereis and Arbacia with special reference to egg extractives. Jour. exp. zoöl. XIV. 1914, VI. The mechanism of fertilization in Arbacia. Ibid., XVI.

—— 1915. Sperm agglutination and fertilization. Biol. bull.,XXVIII.

—— 1921. Studies of fertilization. VIII. On the measure of specificity in fertilization between two associated species of the sea-urchin genus Strongylocentrotus. IX. On the question of superposition of fertilization on parthenogenesis in Strongylocentrotus purpuratus. Biol. bull. XL. X. The effects of copper salts on the fertilization reactions in Arbacia and a comparison of mercury effects. Ibid., XLI.

Lillie, F. R., and E. E. Just. 1924. Fertilization. General cytology. Chicago.

Wilson, E. B. 1928. The cell in development and inheritance. New York.

CHAPTER IV

CLEAVAGE OF THE EGG

Boveri, Th. 1901. Die Polarität von Ovocyte, Ei und Larve des Strongylocentrotus lividus. Zool. Jahrb., XIV.

—— 1902. Über mehrpolige Mitosen als Mittel zur Analyse des Zellkerns. Verh. Phys.-med. Ges. Würzburg. N. F., XXXV.

—— 1905. Über Doppelbefruchtung. Sitzb. Phys.-med. Ges. Würzburg.

Cerfontaine, P. 1906. Recherches sur le développement de l'Amphioxus. Arch. de biol., XXII.

Conklin, E. G. 1905. The organization and cell-lineage of the ascidian egg. Jour. Acad. nat. sci. Phil., XIII.

—— 1905. Mosaic development in ascidian eggs. Jour. exp. zoöl., II.

Driesch, H. 1891. Entwickelungsmechanische Studien. I–II. Zeit. wiss.

Zool. LIII. 1892, III–IV, *Ibid.*, LV. 1893, VII–X, Mitt. Zool. Sta. Neapel, XI.

Driesch, H. 1894. Analytische Theorie der organischen Entwickelung. Leipzig.

——— 1900. Die isolirten Blastomeren des Echinidenkeimes. Arch. Entw-mech., X.

Gray, J. 1927. The mechanism of cell-division. III. The relationship between cell-division and growth in segmenting eggs. Brit. jour. exp. biol., IV.

Just, E. E. 1912. The relation of the first cleavage plane to the entrance point of the sperm. Biol. bull. XXII.

Moore, A. R. 1933. Is cleavage rate a function of the cytoplasm or of the nucleus? Brit. jour. exp. biol., X.

Moore, M. M. 1932. On the coherence of the blastomeres of sea urchin eggs. Arch. Entw.-mech., CXXV.

Morgan, T. H. 1927. Experimental embryology. New York.

Morgan, T. H., and Albert Tyler. 1930. The point of entrance of the spermatozoön in relation to the orientation of the embryo in eggs with spiral cleavage. Biol. bull. LVIII.

Prentiss, C. W. 1915. Embryology. Philadelphia.

Tung, Ti-Chow. 1933. Recherches sur la détermination du plan médian dans l'oeuf de Rana fusca. Arch. de biol., XLIV.

Van Beneden, Ed., and Ch. Julin. 1884. La Segmentation chez les Ascidiens et ses rapports avec l'organisation de la larve. Arch. de biol., V.

Wilson, E. B. 1892. The cell-lineage of Nereis. Jour. morph., VI.

CHAPTER V

GASTRULATION

Lillie, F. R. 1908. The development of the chick. New York.

Patten, B. M. 1929. The early embryology of the chick. 2d ed., Philadelphia.

——— 1931. The embryology of the pig. Philadelphia.

Patterson, J. T. 1909. Gastrulation in the pigeon's egg; a morphological and experimental study. Jour. morph., XX.

——— 1910. Studies on the early development of the hen's egg. 1. History of the early cleavage and of the accessory cleavage. Jour. morph., XXI.

Rhumbler, L. 1902. Zur Mechanik des Gastrulationsvorgange. .... Arch. Entw.-mech., XIV.

Spek, J. 1918. Differenzen im Quellungszustand der Plasmakolloide als eine Ursache der Gastrulainvagination. . . . Kolloidchemische Beihefte. IX.

CHAPTER VI
HALE AND WHOLE EMBRYOS

Chabry, L. 1887. L'Embryologie normale & teratologique des Ascidies simples. Thesis. Paris.

Conklin, E. G. 1906. Does half of an ascidian egg give rise to a whole larva? Arch. Entw.-mech., XXI.

——— 1924. Cellular differentiation. General cytology. Chicago.

——— 1931. The development of centrifuged eggs of ascidians. Jour. exp. zoöl., LX.

——— 1933. The development of isolated and partially separated blastomeres of Amphioxus. Jour. exp. zoöl., LXIV.

Driesch, H. 1898. Resulte und Probleme der Entwickelungsphysiologie der Thiere. Ergeb. anat. Entw., VIII.

——— 1899. Die Lokalisation morphogenetischer Vorgänge. Arch. Entw.-mech., VIII.

——— 1901. Die organischen Regulationen. Leipzig.

——— 1902. Neue Ergänzungen zur Entwickelungsphysiologie des Echinidenkeimes. Arch. Entw.-mech., XIV.

Plough, Harold H. 1927. Defective pluteus larvae from isolated blastomeres of Arbacia and Echinarachnius. Biol. bull., LII.

Roux, W. 1888. Beiträge zur Entwickelungsmechanik des Embryo. 5. Virchow's Archiv., CXIV.

——— 1892. Ueber das entwicklungsmechanische Vermögen jeder der beiden ersten Furchungszellen des Eies. Verh. Anat. Ges., VI.

——— 1895. Gesammelte Abhandlungen über Entwickelungsmechanik der Organismen. II. Leipzig.

——— 1895. Ueber die verschiedene Entwickelung isolirter erster Blastomeren. Arch. Entw.-mech., I.

Schmidt, G. A. 1933. Schnürungs und Durchscheidingversuche am Annurenkeim. Arch. Entw.-mech., CXXIX.

Seidel, F. 1932. Die Potenzen der Furchungskerne im Libelleni und ihre Rolle bei der Aktivierung des Bildungszentrums. Arch. Entw.-mech., CXXVI.

Simons, E. 1932. Verlagerungsversuche zur Untersuchung der Prospectiven Potenz der Blastomeren des Echinodenkeimes. Arch. Entw.-mech., CXXV.

Spemann, H. 1903. Entwicklungsphysiologische Studien am Triton-Ei. III. Arch. Entw.-mech., XVI.

Spemann, H. 1904. Ueber experimentell erzeugte Doppelbildungen mit cyclopischem Defekt. Zool. Jahrb. Suppl., VII.

―――― 1918. Ueber die Determination der ersten Organanlagen des Amphibienembryo, I–IV. Arch. Entw.-mech., XLIII.

CHAPTER VII

THE DEVELOPMENT OF EGG FRAGMENTS

Boveri, Th. 1886. Ueber die Befruchtungs- und Entwicklungsfähigkeit kernloser Seeigeleier und über die Möglichkeit ihrer Bastardierung. Arch. Entw.-mech., II.

―――― 1903. Ueber den Einfluss der Samenzelle auf die Larvencharaktere der Echiniden. Arch. Entw.-mech., XVI.

―――― 1918. Zwei Fehlerquellen bei Merogonieversuchen und die Entwicklungsfähigkeit merogonischer und partiellmerogonischer Seeigelbastarde. Arch. Entw.-mech., XLIV.

Dalcq, A. 1932. Etude des localisations germinales dans l'oeuf vierge d'Ascidie. Arch. d'anat. micros., XXVIII.

―――― 1932. Expériences de merogone sur l'oeuf d'Ascidiella aspersa. C. r. Soc. biol., CIX.

Driesch, H., and T. H. Morgan. 1895. Zur Analysis der ersten Entwickelungs-stadien des Ctenophoreneies. I–II. Arch. Entw.-mech., II.

Harnly, M. H. 1926. Localization of the micromere material in the cytoplasm of the egg of Arbacia. Jour. exp. zoöl., XLV.

Hörstadius, S. 1928. Ueber die Determination des Keimes bei Echinodermen. Acta zoöl., IX.

Schwalbe, E. 1907. Die Doppelbildung. Jena.

Taylor, C. V., and D. H. Tennent. 1924. Preliminary report on the development of egg fragments. Carnegie Inst. Wash., Year book, No. 23.

Tennent, D. H., C. V. Taylor, and D. M. Whitaker. 1929. An investigation on organization in a sea-urchin egg. Carnegie Inst. Wash., Publ. No. 391.

Wilson, E. B. 1903. Experiments on cleavage and localization in the Nemertine-egg. Arch. Entw.-mech., XVI.

―――― 1904. Experimental studies on germinal localization. I, II. Jour. exp. zoöl., I.

Yatsu, N. 1904. Experiments on the development of egg fragments in Cerebratulus. Biol. bull., VI.

Zeleny, C. 1904. Experiments on the localization of developmental factors in the nemertine egg. Jour. exp. zoöl., I.

CHAPTER VIII

SINGLE EMBRYO FROM TWO EGGS

Balinsky, B. 1932. Interaction of two heteropolar equipotential systems .... Jour. bio-zoöl., Acad. sci., Ukraine.

Bierens de Haan, J. A. 1913. Ueber die Entwicklung heterogener Verschmelzungen bei Echiniden. Arch. Entw.-mech., XXXVII.

Kautzsch, G. 1913. Studien über Entwicklungsanomalien bei Ascaris. I. Arch. Zellf., II.

Mangold, O. 1920. Fragen der Regulation und Determination an umgeordneten Furchungsstadien und verschmolzenen Keimen von Triton. Arch. Entw.-mech., XLVII.

Nusbaum, J., and M. Oxner. 1914. Doppelbildungen bei den Nemertinen. Arch. Entw.-mech., XXXIX.

Strassen, O. zur. 1898. Ueber die Riesenbildung bei Ascaris-Eiern. Arch. Entw.-mech., VII.

CHAPTER IX

TWINS AND TWINNING

Conklin, E. G. 1933. The development of isolated and partially separated blastomeres of Amphioxus. Jour. exp. zoöl., LXIV.

Galton, F. 1883. Inquiries into human faculty and its development. New York.

——— 1889. Natural inheritance. London.

——— 1892. Hereditary genius. London.

Muller, H. J. 1925. Mental traits and heredity. Jour. heredity, XVI.

Newman, H. H. 1917. The biology of twins. Chicago.

——— 1923. The physiology of twinning. Chicago.

Patterson, J. T. 1913. Polyembryonic development in Tatus a novemcincta. Jour. morph., XXIV.

Rauber, A. 1880. Formbildung und Formstörung n der Entwicklung von Wirbelthieren. Morph. Jahrb., VI.

Schleip, W., and A. Penners. 1925. Ueber die Duplicitus cruciata bei den O. Schultze'schen Doppelbildungen von Rana fusca. Verh. phys.-med. Ges. Würzburg, L.

——— 1925. Zur Kenntnis der O. Schultzeschen Doppelbildungen beim Frosch. Verh. Phys.-med. Ges. Würzburg, L.

——— 1926. Weitere Untersuchungen über die Entstehung der Schultzeschen Doppelbildungen beim braunen Frosch. Verh. Phys.-med., Ges. Würzburg, LI.

Schwalbe, E. 1907. Die Doppelbildungen. Jena.

Spemann, H. 1916. Ueber transplantation an Amphibienembryonen im Gastrulastadium. Sitzb. Ges. naturforsch. Freunde. Berlin, IX.

―――― 1919. Experimentelle Forschung zum Determinations und Individualitätsproblem. Naturwissenschaften, VII.

―――― 1920. Mikrochirurgische Operationstechnic. Handb. biol. Arbeitsmeth., III.

Stockard, C. R. 1921. Developmental rate and structural expression: An experimental study of twins, double monsters and single deformities. Amer. jour. anat., XXVIII.

Tannreuter, G. W. 1919. Partial and complete duplicity in chick embryos. Anat. rec., XVI.

Tyler, A. 1930. Experimental production of double embryos in annelids and molluscs. Jour. exp. zoöl., LVII.

Wetzel, G. 1895. Über die Bedeutung der cirkulären Furche in der Entwicklung der Schultze'schen Doppelbildungen von Rana fusca. Arch. mikr. Anat., XLVI.

―――― 1896. Beitrag zum Studien kunstlichen Doppelmissbildungen bei Rana fusca. Dissertation. Berlin.

Wilder, H. H. 1904. Duplicate twins and double monsters. Amer. jour. anat., III.

Wilson, E. B. 1892. On multiple and partial development in Amphioxus. Anat. Anz., VII.

―――― 1893. Amphioxus and the mosaic theory of development. Jour. morph., VIII.

CHAPTER X

MULTIPLE CHROMOSOME TYPES

Blakeslee, A. F. 1921. Types of mutations and their possible significance in evolution. Amer. nat., LV.

―――― 1922. Variations in Datura, due to changes in chromosome number. Amer. nat., LVI.

―――― 1924. Distinction between primary and secondary chromosomal mutants in Datura. Proc. Nat. acad. sci., X.

Blakeslee, A. F., and J. Belling. 1924. Chromosomal mutations in the jimson weed, Datura stramonium. Jour. heredity, XV.

Bridges, C. B. 1921. Triploid intersexes in Drosophila melanogaster. Science, LIV.

―――― 1925. Haploidy in Drosophila melanogaster. Proc. Nat. acad. sci., XI.

Marchal, Él., and Ém. Marchal. 1906. Recherches expérimentales sur la sexualité des spores chez les mousses dioïques. Mém. couronnés par la Classe des sciences, dans la séance du 15 décembre 1905.

Marchal, El., and Ém Marchal. 1907, 1911 and 1919. Aposporie et sexualité chez les mousses. Bull. de l'Acad. roy. de Belg. (Classe de science) Nos. 7, 9–10, 1.

Meves, F. 1907. Die Spermatocytenteilungen bei der Honigbiene. Arch. mikr. Anat. LXX.

Morgan, T. H. 1928. The theory of the gene. New Haven.

Schrader, F. 1923. Haploidie bei einer Spinnmilbe. Arch. mikr. Anat., XCVII.

———— 1929. Experimental and cytological investigations of the life-cycle of Gossyparia spuria (Coccidae) and their bearing on the problem of haploidy in males. Zeit. wiss. Zool., CXXXIV.

Schrader, F., and S. Hughes-Schrader. 1926. Haploidy in Icerya purchasi. Zeit. wiss. Zool., CXXVIII.

———— 1931. Haploidy in Metazoa. Quart. rev. biol., VI.

Shull, A. F. 1921. Chromosomes and the life cycle of Hydatina senta. Biol. bull. XLI.

Whiting, P. W. 1921. Heredity in wasps. The study of heredity in a parthenogenetic insect, the parasitic wasp, Hadrobracon. Jour. heredity, XII.

Whitney, D. D. 1909. Observations on the maturation stages of parthenogenetic and sexual eggs of Hydatina senta. Jour. exp. zoöl., VI.

———— 1914. The influence of food in controlling sex in Hydatina senta. Jour. exp. zoöl., XVII.

———— 1916. The control of sex by food in five species of rotifers. Jour. exp. zoöl., XX.

———— 1917. The relative influence of food and oxygen in controlling sex in rotifers. Jour. exp. zoöl., XXIV.

———— 1918. Further studies on the production of functional and rudimentary spermatozoa in rotifers. Biol. bull., XXXIV.

———— 1924. The chromosome cycle in the rotifer Asplanchna intermedia. Anat. rec., XXIX.

CHAPTER XI

PROTOPLASM AND GENES

Baltzer, F. 1909. Die Chromosomen von Strongylocentrotus lividus und Echinus microtuberculatus. Arch. Zellf., II.

———— 1910. Ueber die Entwicklung der Echiniden-Bastarde mit besonderer Berücksichtigung der Chromatinverhältnisse. Zool. Anz., XXXV.

———— 1910. Ueber die Beziehung zwischen dem Chromatin und der Entwicklung und Vererbungsrichtung bei Echinodermenbastarden. Arch. Zellf., V.

Boveri, Th. 1904. Noch ein Wort über Seeigelbastarde. Arch. Entw.-mech., XVII.

────── 1914. Ueber die Charaktere yon Echiniden-Bastardlarven bei verschiedenem Mengenverhältnis mütterlicher und väterlicher Substanzen. Verh. Phys. med. Ges. Würzburg, XLIII.

Boycott, A. E., and C. Diver. 1923. On the inheritance of sinistrality in Limnaea peregra. Proc. Roy. soc. London, B, XCV.

Boycott, A. E., C. Diver, S. L. Garstang, and F. M. Turner. 1930. The inheritance of sinistrality in Limnaea peregra. Proc. Roy. soc. London, B, CCIX.

Chambers, R. 1924. The physical structure of protoplasm as determined by micro-dissection and injection. General cytology. Chicago.

Clausen, R. E., and M. C. Mann. 1924. Inheritance in Nicotiana tabacum, V. The occurrence of haploid plants in interspecific crosses. Proc. Nat. acad. sci., X.

Crampton, H. E. 1894. Reversal of cleavage in a sinistral gastropod. Ann. N. Y. acad. sci., VIII.

Driesch, H. 1903. Ueber Seeigelbastarde. Arch. Entw.-mech., XVI.

Emerson, S. H. 1929. The reduction division in a haploid Oenothera. La Cellule, XXXIX.

Fischel, A. 1906. Ueber Bastardierungsversuche bei Echinodermen. Arch. Entw.-mech., XXII.

Hayes, H. K., and E. M. East. 1915. Further experiments on inheritance in maize. Conn. agri. exp. sta. bull., CLXXXVIII.

Heilbrunn, L. V. 1928. The colloid chemistry of protoplasm. Berlin.

Herbst, C. 1906. Vererbungsstudien. I–III. Arch. Entw.-mech., XXI–II; 1907, V. *Ibid.*, XXIV; 1909, VI. *Ibid.*, XXVII; 1912, VII. *Ibid.*, XXXIV.

────── 1913. Vererbungsstudien. VIII, IX. Sitzb. Heidelberger Akad. Wiss.

Hertwig, G., and P. Hertwig. 1914. Kreuzungsversuche an Knochenfischen. Arch. mikr. Anat., LXXXIV.

Koehler, O. 1915. Ueber die Ursachen der Variabilität bei Gattungsbastarden von Echiniden. I, II. Zeit. abst. Vererb., XV.

Loeb, J. 1904. Further experiments on the fertilization of the egg of the sea urchin with sperm of various species of starfish and a holothurian. Univ. Calif. publ. physiol., I.

────── 1906. The dynamics of living matter. Columbia Univ., Biol. series.

────── 1908. Ueber die Natur der Bastardlarve zwischen dem Echinodermenei (Strongylocentrotus franciscanus) und Molluskensamen (Chlorostoma funebrale). Arch. Entw.-mech., XXVI.

Loeb, J. 1912. Heredity in heterogeneous hybrids. Jour. morph., XXIII.

Moenkhaus, W. J. 1910. Cross-fertilization among fishes. Proc. Ind. acad. sci.

Nachtsheim, H. 1921. Sind haploide Organismen lebensfähig? Biol. Zentrlb., XLI.

Newman, H. H. 1914. Modes of inheritance in teleost hybrids. Jour. exp. zoöl., XVI.

—— 1915. Development and heredity in heterogenic teleost hybrids. Jour. exp. zoöl., XVIII.

Pinney, Edith. 1918. A study of the relation of the behavior of the chromatin to development and heredity in teleost hybrids. Jour. morph., XXXI.

Renner, O. 1924. Die Scheckung der Oenotherenbastarde. Biol. Zentrlb., XLIV.

Schrader, F., and S. Hughes-Schrader. 1931. Haploidy in Metazoa. Quart. rev. biol., VI.

Sturtevant, A. H. 1923. Inheritance of direction of coiling in Limnaea. Science, LVIII.

Tennent, D. H. 1908. The chromosomes in cross-fertilized Echinoid eggs. Biol. bull., XV.

—— 1910. The dominance of maternal or of paternal characters in Echinoderm hybrids. Arch. Entw.-mech., XXIX.

—— 1910. Variation in Echinoid plutei. Jour. exp. zoöl., IX.

—— 1911. Echinoderm hybridization. Carnegie Inst. Wash., Publ. No. 132.

—— 1923. Investigations on the hybridization of Echinoids. Carnegie Inst., Wash., Year book, No. 22.

—— 1925. Investigations on the specificity of fertilization. Carnegie Inst. Wash., Year book, No. 24.

Toyama, K. 1912. On certain characteristics of the silkworm apparently non-mendelian. Biol. Centralb., XXXII.

Wettstein, F. von. 1924. Morphologie und Physiologie des Formwechsels der Moose auf genetischer Grundlage. I. Zeit. abst. Vererb., XXXIII.

—— 1924. Ueber Fragen der Geschlechtsbestimmung bei Pflanzen. Die Naturwissenschaften, XXXVIII.

—— 1928. Über plasmatische Vererbung und über das Zusammenwirken von Genen und Plasma. Berich. Deutschen bot. Ges., XLVI.

—— 1930. Über plasmatische Vererbung, sowie Plasma- und Genwirkung, II. Ges. Wiss. Göttingen, Biologie, VI.

CHAPTER XII

LARVAL AND FOETAL TYPES

Baer, K. E. von. 1828. Über Entwicklungsgeschichte der Thiere. Königsberg.

Bolk, L. 1926. Das Problem der Menschwerdung. Jena.

Garstang, W. 1922. The theory of recapitulation. Jour. Linnaean society.

———— 1928. The origin and evolution of larval forms. Rep. Brit. assoc.

Haeckel, E. 1866. Generelle Morphologie der Organismen. Berlin.

Herdman, W. A. 1904. Ascidians and Amphioxus. Cambridge natural history, VII.

Seeliger, O. 1885. Die Entwicklungsgeschichte der socialen Ascidian. Jena Zeit. Naturwissenschaft, XVIII.

CHAPTER XIII

PARTHENOGENESIS

Artom, C. 1924. Il tetraploidismo dei maschi dell' Artemia salina di Odessa in relazione con alcuni problemi generali di genetica. Rend. Reale accad. naz. Lincei, XXXII.

———— 1930. L'origine e l'evoluzione della partenogenesi nell' Artemia salina diploide di Sète. Rend. Reale accad. naz. Lincei, XI (6).

Baehr, W. B. von. 1920. Recherches sur la maturation des oeufs parthénogénétiques dans l'Aphis Palmae. La Cellule, XXX.

Banta, A. M., and L. A. Brown. 1929. Control of sex in Cladocera. Proc. Nat. acad. sci., XV.

Banta, A. M., and T. R. Wood. 1928. Inheritance in parthenogenesis and in sexual reproduction in Cladocera. Zeit. abst. Vererb., Suppl. I.

Bataillon, E. 1910. L'Embryogénèse complète provoquée chez les Amphibiens par piqûre de l'oeuf vierge, larves parthénogénétiques de Rana fusca. C. R. acad. sci., CL.

Dalcq, A. 1928. Les bases physiologiques de la fécondation et de la parthénogénèse. Paris.

Herlant, M. 1913. Étude sur les bases cytologiques du mécanisme de la parthénogénèse expérimentale chez les Amphibiens. Arch. de biol., XXVIII.

Loeb, J. 1913. Artificial parthenogenesis and fertilization. Chicago.

———— 1920–21. Further observations on the production of parthenogenetic frogs. Jour. gen. physiol., III.

Loeb, J., and F. W. Bancroft. 1912. Can the spermatozoön develop outside the egg? Jour. exp. zoöl., XII.

Morgan, T. H. 1912. The elimination of the sex-chromosomes from the male-producing eggs of Phylloxerans. Jour. exp. zoöl., XII.

—— 1915. The predetermination of sex in Phylloxerans and Aphids. Jour. exp. zoöl., XIX.

Parmenter, C. L. 1920. The chromosomes of parthenogenetic frogs. Jour. gen. physiol., II.

—— 1925. The chromosomes of parthenogenetic frogs and tadpoles. Jour. gen. physiol., VIII.

—— 1926. The chromosomes of parthenogentically developed young tadpoles and early cleavages of Rana pipiens. Anat. rec., XXXIV.

Patterson, J. T. 1928. Sexes in the Cynipidae and male-producing and female-producing lines. Biol. bull., LIV.

Shull, A. F. 1910. Studies in the life cycles of Hydatina senta. Jour. exp. zoöl., VIII.

—— 1915. Inheritance in Hydatina senta. Jour. exp. zoöl., XVIII.

—— 1915. Periodicity in the production of males in Hydatina senta. Biol. bull., XXVIII.

—— 1928. Duration of light and the wings of the aphid Macrosiphum solanifolii. Arch. Entw.-mech., CXIII.

—— 1929. The effect of intensity and duration of light and of duration of darkness, partly modified by temperature, upon wing-production in aphids. Arch. Entw.-mech., CXV.

—— 1929. Determination of types of individuals in aphids, rotifers and cladocera. Biol. rev., IV.

—— 1930. Control of gamic and parthenogenetic reproduction in winged aphids by temperature and light. Zeit. abst. Vererb., LV.

—— 1930. Order of embryonic determination of the differential features of gamic and parthenogenetic aphids. Zeit. abst. Vererb., LVII.

Tyler, Albert. 1931. The production of normal embryos by artificial parthenogenesis in the echiuroid, Urechis. Biol. bull., LX.

Vandel, A. 1928. La Parthénogénèse géographique. Bull. biol. France et Belg., LXII.

—— 1931. La Parthénogénèse. Paris.

Weismann, A. 1887. Ueber die Zahl der Richtungskörper und über ihre Bedeutung für die Vererbung. Jena.

Weismann, A., and G. Ischikawa. 1889. Weitere Untersuchungen zum Zahlengesetz der Richtungskörper. Zoöl. Jahrb. Abst. anat. ontog., III.

Whitney, D. D. 1914. The influence of food in controlling sex in Hydatina senta. Jour. exp. zoöl., XVII.

—— 1916. The control of sex by food in five species of rotifers. Jour. exp. zoöl., XX.

Whitney, D. D. 1917. The relative influence of food and oxygen in controlling sex in rotifers. Jour. exp. zoöl., XXIV.

—— 1917–18. The production of functional and rudimentary spermatozoa in rotifers. Biol. bull., XXXIII, XXXIV.

—— 1929. The chromosome cycle in the rotifer Asplanchna amorphora. Jour. morph. and phys., XLVII.

Winkler, H. 1920. Verbreitung und Ursache der Parthenogenesis im Plantzen- und Tierreiche. Jena.

CHAPTER XIV

REGENERATION

Bonner, J. 1932. The production of growth substance by Rhizopus suinus. Biol. Zentrlb., LII.

Buy, H. G. du. 1931. Über die Bedingungen, welche die Wuchsstoffproduktion beeinflussen. Proc. Kön. Akad. wetensch. Amsterd., XXXIV.

Buy, H. G. du, und E. Nuernbergk. 1932. Phototropismus und Wachstum der Pflanzen. Ergeb. der Biol., IX.

Child, C. M. 1907. An analysis of form regulation in Tubularia. I–VI. Arch. Entw.-mech., XXIII, XXIV.

Cholodny, N. 1927. Wuchshormone und Tropismen bei den Pflanzen. Biol. Zentrlb., XLVII.

—— 1928. Beiträge zur hormonalen Theorie von Tropismen. Planta, VI.

Dolk, H. E. 1929. Geotropie en groeistof. Dissertation. Utrecht (1930). See also: Über die Wirkung der Schwerkraft auf Koleoptilen von Avena satina. I, II. Proc. Kon. Akad. wetensch. Amsterd., XXXII.

Dolk, H. E., and K. V. Thimann. 1932. Studies on the growth hormone of plants. I. Proc. Nat. acad. sci., XVIII.

Huxley, J. S., and F. S. Callow. 1933. A note on the asymmetry of male fiddler crabs. Arch. Entw.-mech., CXXIX.

Korschelt, E., and K. Heider. 1902. Lehrbuch der vergleichenden Entwicklungsgeschichte der wirbellosen Thiere. Jena.

Lund, E. J. 1921, 1925. Experimental control of organic polarity by the electric current. Jour. exp. zoöl., XXXIV, XLI.

Morgan, T. H. 1901. Regeneration. New York.

—— 1901. Regeneration in Tubularia. Arch. Entw.-mech., XI.

—— 1923. The development of asymmetry in the fiddler crab. Amer. Nat., LVII.

Morgan, T. H., and S. E. Davis. 1902. The internal factors in the regeneration of the tail of the tadpole. Arch. Entw.-mech., XV.

Peebles, F. 1900. Experiments in regeneration and in grafting of Hydrozoa. Arch. Entw.-mech., X.

———— 1902. Further experiments in regeneration and grafting of Hydroids. Arch. Entw.-mech., XIV.

Przibram, H. 1896. Regeneration bei den niederen Crustaceen. Zool. Anz., No. 514.

———— 1899. Die Regeneration bei den Crustaceen. Arb. Zool. Inst. Wien, XI.

———— 1901–2. Experimentelle Studien über Regeneration. I–II. Arch. Entw.-mech., XI, XIII.

Stockard, C. R. 1907. Studies of tissue growth. I. An experimental study of the rate of regeneration in Cassiopea xamachana (Bigelow.) Carnegie Inst. Wash., Publ. No. 103.

———— 1909. II. Functional activity, form regulation, level of cut, and degree of injury as factors in determining the rate of regeneration, the reaction of regenerating tissue in the old body. Jour. exp. zoöl., VI.

Thimann, K. V., and J. Bonner. 1932. Studies on the growth hormone of plants. II. The entry of growth substance into the plant. Proc. Nat. acad. sci., XVIII.

———— 1933. The mechanism of the action of the growth substance of plants. Proc. Roy. soc. London, B, CXIII.

Went, F. W. 1928. Wuchsstoff und Wachstum. Rec. trav. bot. néerl., XXV. Amsterdam.

Wilson, E. B. 1903. Notes on the reversal of asymmetry in the regeneration of the chelae in Alpheus heterochelis. Biol. bull., IV.

Woerdeman, M. W. 1932. Ontogénèse et Regeneration. I, II. Ann. Soc. roy. sci. med. nat. Bruxelles, Bull. 3–4, 5–6.

Wolff, G. 1893. Entwicklungphysiologische Studien. I. Die Regeneration der Urodelenlinse. Arch. Entw.-mech., I.

Zeleny, C. 1904. Compensatory regulation. Jour. exp. zoöl., II.

CHAPTER XV

LOCALIZATION AND INDUCTION

Balinsky, B. I. 1925. Transplantation des Ohrbläschens bei Triton. Arch. Entw.-mech., CV.

———— 1926. Weiteres zur Frage der experimentellen Induktion einer Extremitätenanlage. Arch. Entw.-mech., CVII.

———— 1927. Xenoplastische Ohrbläschentransplantation zur Frage der Induktion einer Extremitätenanlage. Arch. Entw.-mech., CX.

———— 1927. Über experimentelle Induktion der Extremitätenanlage bei Triton mit besonderer Berücksichtigung der Innervation und Symmetrieverhältnisse derselben. Arch. Entw.-mech., CX.

250   REFERENCES

Ekman, G. 1913. Experimentelle Untersuchungen über die Entwicklung der Kiemenregion (Kiemenfäden und Kiemenspalten) einiger anuren Amphibien. Morph. Jahrb., XLVII.

—— 1914. Experimentelle Untersuchungen über die Entwicklung des Peribranchialraumes bei Bombinator. Finska Vet.-Soc. Förhandl., LVI.

Filatow, D. 1916. Exstirpation und Verpflanzung von Gehörbläschen bei Bufo-Larven. Rev. zool. russe, I.

Geinitz, B. 1925. Zur weiteren Analyse des Organisationszentrums. Zeit. abst. Vererb., XXXVII.

Goerttler, K. 1925. Die Formbildung der Medullaranlage bei Urodelen. Arch. Entw.-mech., CVI.

Harrison, R. G. 1925. The effect of reversing the medio-lateral or transverse axis of the fore-limb bud in the salamander embryo. (Amblystoma punctatum Linn.) Arch. Entw.-mech., CVI.

Holtfreter, J. 1933. Nachweis der Induktionsfähigkeit abgetöteter Keimteile. Arch. Entw.-mech., CXXVIII.

—— 1933. Die totale Exogastrulation ... Arch. Entw.-mech., CXXIX.

Hörstadius, S. 1928. Ueber die Determination des Keimes bei Echinodermen. Acta zoöl., IX.

Lewis, W. H. 1907. Transplantations of the lips of the blastopore in Rana palustris. Amer. jour. anat., VII.

Mangold, O. 1920. Fragen der Regulation und Determination an umgeordneten Furchungsstudien und verschmolzenen Keimen von Triton. Arch. Entw.-mech., XLVII.

—— 1924. Transplantationsversuche zur Frage der Spezifität und der Bildung der Keimblätter. Arch. mikr. Anat., C.

—— 1926. Über formative Reize in der Entwicklung der Amphibien. Die Naturwissenschaften, XIV.

—— 1929. Experimente zur Analyse der Determination und Induktion der Medullarplatte. Arch. Entw.-mech., CXVII.

Seevers, C. H., and D. A. Spencer. 1932. Autoplastic transplantation of guinea-pig skin between regions with different characters. Amer. nat., LXVI.

Spemann, H. 1910. Die Entwicklung des invertierten Hörgrübchens zum Labyrinth. Arch. Entw.-mech., XXX, pt. 2.

—— 1918. Ueber die Determination der ersten Organanlagen des Amphibienembryos. I–VI. Arch. Entw.-mech., XLIII.

—— 1925. Some factors of anir..l development. Brit. jour. exp. biol., II.

Streeter, G. L. 1907. Some factors in the development of the Amphibian ear vesicle and further experiments on equilibration. our. exp. zoöl., IV.

Streeter, G. L. 1914. Experimental evidence concerning the determination of posture of the membranous labyrinth in Amphibian embryos. Jour. exp. zoöl., XVI.

Tandler, J., and K. Keller. 1910. Ueber den Einfluss der Kastration auf den Organismus. IV. Die Körperform der weiblichen Frühkastraten des Rindes. Arch. Entw.-mech., XXXI.

CHAPTER XVI
THE DETERMINATION OF SEX

Baltzer, F. 1925. Untersuchungen über die Entwicklung und Geschlechtsbestimmung der Bonellia. Publ. Staz. zool., Napoli, VI.

——— 1926. Über die Vermännlichung indifferenter Bonellia-Larven durch Bonellia-Extrakte. Rev. suisse zool., XXXIII.

——— 1928. Über metagame Geschlechtsbestimmung und ihre Beziehung zu einigen Problemen der Entwicklungsmechanik und Vererbung. Verhandl. Deutschen zoolog. Ges., V.

——— 1928. Neue Versuche über die Bestimmung des Geschlechts bei Bonellia viridis. Rev. suisse zool., XXXV.

Benoit, M. J. 1929–30. Le Déterminisme des caractères sexuals secondaires du coq domestique. Arch. de zool., LXIX.

Bridges, C. B. 1925. Sex in relation to chromosomes and genes. Amer. nat., LIX.

——— 1932. The genetics of sex in Drosophila. Sex and Internal Secretions. Baltimore.

Burns, R. K. 1925. The sex of parabiotic twins in Amphibia. Jour. exp. zoöl., XLII.

Champy, C. 1924. Sexualité et hormones. Paris.

Champy, C., and Th. Keller. 1928. Contribution à l'étude des hormones sexuelles femelles. Arch. de morph., XXVII.

Crew, F. A. E. 1923. Studies in intersexuality. I and II. Proc. Roy. soc. London, B, XCV.

——— 1927. Abnormal sexuality in animals. III. Sex reversal. Quart. rev. biol., II.

Danforth, C. H. 1929. Genetic and metabolic sex-differences. The manifestation of a sex-linked trait following skin transplantation. Jour. heredity, XX.

——— 1929. The effect of foreign skin on feather pattern in the common fowl (Gallus domesticus). Arch. Entw.-mech., CXVI.

——— 1932. Interrelation of genic and endocrine factors in sex. Sex and Internal Secretions. Baltimore.

Danforth, C. H., and F. Foster. 1929. Skin transplantation as a means of studying genetic and endocrine factors in the fowl. Jour. exp. zoöl., LII.

Domm, L. V. 1927. New experiments on ovariotomy and the problem of sex inversion in the fowl. Jour. exp. zoöl., XLVIII.

Domm, L. V., Mary Juhn, and R. G. Gustavson. 1932. Plumage tests in birds. Sex and Internal Secretions. Baltimore.

Friess, E. 1933. Untersuchungen über die Geschlechtsumkehr bei Xiphiphorus Helleri. Arch. Entw.-mech., CXXIX.

Goldschmidt, R. 1931. Die Sexuellen Zwischenstuffen. Berlin.

Goldschmidt, R., and K. Katsuki. 1927. Erblicher Gynandromorphismus und somatische Mosaikbildung bei Bombyx mori L. Biol. Zentrlb. XLVII.

——— 1928. Zweite Mitteilung über erblichen Gynandromorphismus bei Bombyx mori L. Biol. Zentrlb., XLVIII.

——— 1928. Cytologie des erblichen Gynandromorphismus von Bombyx mori L. Biol. Zentrlb., XLVIII.

Greenwood, A. W. 1928. Studies on the relation of gonadic structure to plumage characterisation in the domestic fowl. IV. Gonad cross transplantation in Leghorn and Campine. Proc. Roy. soc. London, B., CIII.

Greenwood, A. W., and J. S. S. Blyth. 1929. An experimental analysis of the plumage of the Brown-Leghorn fowl. Proc. Roy. soc. Edinburgh, XLIX.

Guyénot, E., and K. Ponse. 1927. Questions théoretiques soulevées par le cas de l'organe de Bidder du crapaud. C. R. soc. biol. Paris, XCVI.

Harms, J. W. 1923. Untersuchungen über das Biddersche Organ der männlichen und weiblichen Kröten. II. Zeit. Anat., LXIX.

——— 1926. Beobachtungen über Geschlechtsumwandlungen reifer Tiere und deren F_2-Generation. Zool. Anz., LXVII.

——— 1926. Körper- und Keimzellen. I–II. Monographien physiol., IX. Berlin.

Herbst, C. 1928. Untersuchungen zur Bestimmung des Geschlechts. I. Ein neuer Weg zur Lösung des Geschlechts-bestimmungsproblems bei Bonellia viridis. Sitz. Heidelberger Akad. Wiss., 1928.

——— 1929. II. Weitere Experimente über die Vermännlichung indifferenter Bonellia-Larven durch künstliche Mittel. Sitz. Heidelberger Akad. Wiss., 1929.

Junker, H. 1923. Cytologische untersuchungen an den Geschlechtsorganen der halbzwitterigen Steinfliege Perla marginata. Arch. Zellf., XVII.

Kahle, W. 1908. Die Paedogenesis der Cecidomyiden. Stuttgart.

Lillie, F. R. 1917. The freemartin: a study of the action of sex hormones in the foetal life of cattle. Jour. exp. zoöl., XXIII.

———— 1927. The present status of the problem of sex-inversion in the hen. Jour. exp. zoöl., XLVIII.

Meisenheimer, J. 1909. Über den Zusammenhang primärer und sekundärer Geschlechtsmerkmale bei den Schmetterlingen und den übrigen Gliedertieren. Exp. Stud. z. Soma-u. Geschlechtsdiff, I. Jena.

———— 1921. Geschlecht und Geschlechter im Tierreiche. I. Die natürlichen Beziehungen. Jena.

———— 1930. II. Die allgemeinen Probleme. Jena.

Pézard, A. 1918. Le Conditionnement physiologique des caractères sexuels secondaires chez les oiseaux. Bull. sci. France et Belge, LII.

Ponse, K. 1924. L'Organe de Bidder et le déterminisme des caractères sexuels secondaires du crapaud (Bufo vulgaris). Rev. suisse zool., XXXI.

———— 1925. Ponte et développement d'oeufs provenant de l'orgar.e de Bidder d'un crapaud male féminisé. C. R. soc. biol. Paris, XCII.

———— 1927. Les potentialités de l'organe de Bidder des crapauds. C. R. soc. biol., Paris, XCVI.

———— 1927. Les Hypothèses concernant la signification de l'organe de Bidder du crapaud. C. R. soc. biol., Paris, XCVI.

Schrader, F. 1920. Sex determination in the white fly (Trialeurodes vaporariorum). Jour. morph., XXXIV.

———— 1928. Die Geschlechtschromosomen. Berlin.

Steinach, E. 1912. Willkürliche Umwandlung von Säugetier-Männchen in Tiere mit ausgeprägt weiblichen Geschlechtscharakteren und weiblicher Psyche. Arch. Physiol., CXLIV.

Stern, C. 1927. Ein genetischer und zytologischer Beweis für Vererbung im Y-chromosome von Drosophila melanogaster. Zeit. abst. Vererb., XLIV.

Tandler, J. 1910. Über den Einfluss der Geschlechtsdrüsen auf die Geweihbildung bei Rentieren. Anz. Akad. Wiss. Wien, Math.-naturwiss. Kl., XLVII.

Whiting, P. W. 1918. Sex determ'nation and biology of a parasitic wasp, Hadrobracon brevicornis. Biol. bull., XXXIV.

Willier, B. H. 1924. The endocrine glands and the development of the chick. Amer. jour. anat., XXXIII.

———— 1932. Embryological foundations of sex in Vertebrates. Sex and Internal Secretions. Baltimore.

Witschi, E. 1925. Studien über Geschlechtsumkehr und sekundäre Geschlechtsmerkmale der Amphibien. Arch. d. Julius Klaus-Stiftung. I.

Witschi, E. 1927. Sex-reversal in parabiotic twins of the American Wood-frog. Biol. bull., LII.

———— 1928. Effect of high temperature on the gonads of frog larvae. Proc. Soc. exp. biol. med., XXV.

———— 1932. Sex deviations, inversions, and parabiosis. ex and Internal Secretions. Baltimore.

Zawadowsky, M. M. 1922. Das Geschlecht und die Entwicklung der Geschlechtsmerkmale. Moscow, Staatsverlag.

———— 1926. Bisexual nature of the hen and experimental hermaphroditism in hens. Trans. Lab. exp. biol., Zoopark, Moscow, II.

———— 1929. Die Äquipotentialität der Gewebe des Männchens und Weibchens bei Vögeln und Säugetieren. Endokrinologie, V.

CHAPTER XVII

PHYSIOLOGICAL EMBRYOLOGY

Conklin, E. G. 1929. Problems of development. Amer. nat., LXIII.

Morgan, T. H. 1926. Genetics and the physiology of development. Amer. nat., LX.

Needham, Joseph. 1926. The energy sources in ontogenesis. I, II, III. Jour. exp. biol., III, IV; 1927, IV, V, VI. *Ibid.*, IV, V; 1933, VII. *Ibid.*, X.

———— 1930. The biochemical aspects of the recapitulation theory. Biol. rev., V.

———— 1931. Chemical embryology, Vols. I–III. Cambridge.

———— 1933. On the dissociability of the fundamental processes in ontogenesis. Biol. rev., VIII.

Warburg, O. 1928. Ueber die katalytischen Wirkungen der lebendigen Substanz. Berlin.

INDEX